Brooklands Books

CW01667714

T SERIES RESTORATION GUIDE

By Malcolm Green

Brooklands Books

BROOKLANDS BOOKS LTD.
P.O. BOX 146, COBHAM,
SURREY, KT11 1LG. UK
sales@brooklands-books.com

www.brooklands-books.com

MGTR

Printed in China

Published by

Brooklands Books

Brooklands Books Limited
P.O. Box 146, Cobham, Surrey, KT11 1LG England.
Tel: 01932 865051 Fax: 01932 868803
e-mail: sales@brooklands-books.com

ISBN 9781855202115
MGTR

First Published 1993
Second revised and enlarged edition 2000
Third Edition 2003
Fourth Edition 2007
Fifth Edition 2011

Whilst every care has been taken in the preparation of this book, neither the publisher nor the author can accept any liability for loss, damage or injury caused by errors in, or omissions from, the information given. Readers are reminded that any work they carry out, or modifications they make to vehicles, as a result of following the guidance given here are done at their risk and they should satisfy themselves of the safety and legality of any such operations before proceeding, and if in doubt take professional advice.

www.brooklands-books.com

Printed and bound in China

TABLE OF CONTENTS

ACKNOWLEDGEMENTS

Although I had been running M.G.s as everyday cars for some years it was not until 1969 that I first discovered the pleasure that owning a T-type could bring when I purchased a 1937 TA as a box of bits for £170. Rebuilding this and subsequent cars brought me in touch with other owners all over the world and many of the friendships made then have lasted to the present day.

When I first started to work on the car I knew very little and leant heavily on advice and help from other M.G. Car Club members. Later, having been introduced to the New England M.G. T Register by its enthusiastic chairman Dick Knudson, I had the benefit of advice given by the contributors to the register journal "The Sacred Octagon". When compiling this book I have drawn, both on my own experiences when rebuilding T-types, and on the articles and books on the subject that have appeared over the years. To the authors of all of these I can only offer my grateful thanks for the help they have unknowingly provided.

I have not tried to re-write the workshop manuals as copies of the TC and TD/TF manuals are available from Brooklands Books along with the driver's handbooks for the TD and TF and there is little point in repeating the advice they contain. However, some of the drawings and details contained in the TD/TF manual have been included where these would be of assistance to the owners of earlier cars.

The rebuild seminars held by the M.G. Car Club T Register are highly recommended and I have certainly drawn on the words of wisdom of lecturers Mike Collingburn, Ron Gammons and Gerry Brown, who also had the task of helping me with the engine and mechanical sections of this book. I was lucky in being able to also call on the assistance of many people when working on the book and would particularly like to acknowledge the help given by Mike Beales, for allowing me to examine and photograph his TFs, Roger Chamberlain who landed the job of helping with the proof reading of the manuscript, Mike Collingburn who kindly checked the trimming section, Ron Cover for his advice over the years in addition to all the photographs acknowledged below, Dave Jarvis for the help with TA details, John Marks of Vintage Restorations, Dave Raymond for his help including the details of the TC EX-U, and Roger Thomas who found the time to illustrate the engine rebuilding section.

For more than thirty years the world's leading motoring journals have allowed Brooklands Books to re-issue their copyright material and I gratefully acknowledge their generosity in allowing the reproduction here of some drawings and illustrations from Autocar, Light Car, Motor and Motor Trader. I am also grateful to the British Motor Industry Heritage Trust for the use of drawings from the workshop manuals.

PHOTOGRAPH CREDITS

All photographs by the author, or from his collection, except for the following:-
Mike Collingburn 121 top, Ron Cover 38 top, 44, 55 top and centre, 56,61, Richard Monk 120 top, Richmond Pike 12.
It has not been possible in every case to trace the original photographer and if any copyright has been infringed it has been done unintentionally and the sincerest of apologies are offered.

CHAPTER ONE

INTRODUCTION

The initials M.G. will mean different things to different generations. If you are a child of the seventies or eighties the first image visualised could well be that of an MGB or even of an M.G. Metro. However, for those brought up in earlier years I am sure the the sight and sound of a T-series M.G. being enjoyed to the full would spring to mind.

I know that the exploits of the earlier four and six cylinder cars produced between 1929 and 1936 played a major part in establishing the charisma of the M.G. sports car. For a time in the 1930's the reputation of the British motor industry abroad rested almost solely on the output from this small Abingdon factory if judged on competition results. However, for the average enthusiast in the post-war years, it was those sturdy little cars produced between 1936 and 1955 that gave them their first taste of sports car driving. Indeed, if you study the careers of a large number of well known competition drivers, you will find that the first experience of their craft was gained driving one of these cars.

It is easy now to dismiss the post-war TC, TD and TF as being old fashioned cars, even when they were new, and rather heavy and slow compared with the opposition. This, though, ignores their basic virtues of strength, both of the chassis and of engine, and comparative cheapness and availability. It is these virtues today that have made them so popular with people now looking for something different to restore and run. Although looking akin to the pre-war cars, the TD and TF are particularly suitable for use everyday, even in modern traffic. They benefit from a fairly rigid chassis and a suspension set up very similar to the later MGA and MGB. The rack and pinion steering gives good control and, with radial tyres, you have the confidence to use the roadholding to the full. That is not to say that a TA, TB or TC is unsuitable for everyday usage, just that I prefer the later cars when the roads are rough and the traffic heavy!

THE DEVELOPMENT OF THE CARS.

The first T-type was the TA. This was designed as part of the rationalisation process after the amalgamation of Morris Motors, Wolseley and M.G. in 1935. The design office at Abingdon was closed and work on all new models carried out at Cowley. The man responsible for the overall design was H. N. Charles and he was able to use the then new O.H.V. engine destined for the Wolseley 10/40, albeit with modifications to the camshaft and manifolding, which, together with twin SU carburettors, increased the power output to around 50 bhp. This engine was mated to a four-speed gearbox driven through a cork faced clutch running in oil.

The new Midget, as it was called, had to replace all the existing sporting O.H.C. cars then produced by

Abingdon. There were, of course, new larger saloons and tourers but these did not really appeal to the man looking for a car to use for competition, or as a replacement for his beloved P or L-type 2-seater. The new T-type Midget was larger than the earlier cars but benefited from a chassis that was developed from one used by them. This was of a simple ladder design with the two main frames set parallel and separated by tubular crossmembers. Those tubular crossframes also formed the mounting points for the springs. These were pivoted at the front end, but at the rear, instead of using conventional shackles, they were mounted into bronze trunnions; a system which was said to stiffen up the chassis and to stop the springs moving laterally. This chassis continued after the war in the TC but the spring mountings were changed to the more conventional shackles. The sliding trunnions were very prone to wear rapidly if not well lubricated and, if these were not then replaced, the chassis cross tube itself would be worn away by the spring bearing directly on it.

As with the earlier cars the TA was used in competition, notably in production car trials, and was occasionally fitted with the larger VA engine. The new Midget was a sales success, with over 3,000 being built, before it was replaced by the short lived TB in 1939. The TB looked identical to the TA it replaced and in most respects it was. The really important change was the introduction of a new, and very sturdy, engine. This power unit, the XPAG, was to remain in use, with some minor and one major change, up to the end of production of the TF in 1955.

This new power unit originally appeared as the XPJM for the Morris 10 in 1938 and it was a completely new design with a shorter 90mm stroke (the TA unit was 102mm) and a much more robust block. The cylinder head design was good with separate exhaust ports for each cylinder to aid gas flow. The whole unit was mated to a gearbox with synchromesh on three of its four gears and was driven through a dry clutch. It was modified for the M.G. by increasing the bore to bring it up to 1,250cc and improving the manifolding and adding twin carburettors. The new engine was very free revving and the rear axle ratio was slightly lowered to take advantage of this. This changed the whole character of the car. The almost "lazy" driving technique that suited the TA, not revving the engine too hard and

The start of it all. The prototype TA.

relying on the good torque produced, is replaced by a more sporting and urgent driving style. I am not saying that the TA was not a good car - merely different.

In addition to the standard 2-seater, from August 1938 first the TA, and later the TB, were available with the attractive Tickford body. This featured a "three position" drop head and full doors that incorporated winding windows. The interior was trimmed to a high standard and was all together more civilised than the standard car. These models were built by the coachbuilders, Salmons and Sons Limited of Newport Pagnell, on completed running chassis supplied from Abingdon. These Tickford bodied cars were built alongside the other Abingdon cars, the VA and SA (later WA), supplied with similarly attractive bodies.

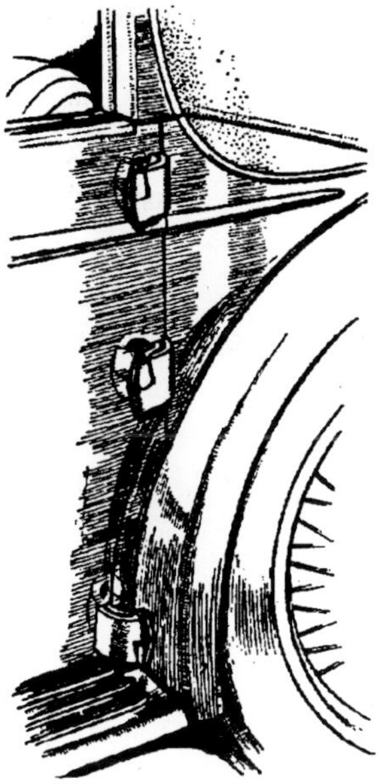

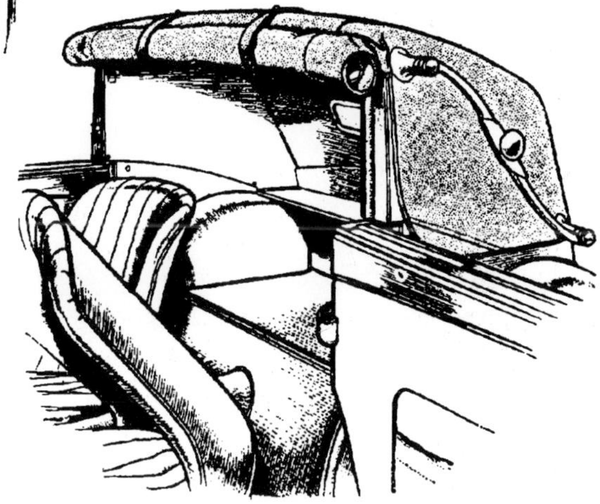

One of the prototype Tickford Coupes.

The TC. This is a late model car with the diamond pattern headlights, tan cover and SFT462 spot lamp.

When Abingdon returned to car production after the war the TB was looked at and modifications to the design made to try to answer criticisms by customers. As recalled earlier they decided to change the chassis to incorporate spring shackles, rather than sliding trunnions, for the rear mountings of the leaf springs. The body was widened slightly which meant that the running boards were narrower. The reduction of the number of tread strips from three to two for the TC is one of the major recognition points to separate the pre-war cars from the post-war at a glance. There were also changes to the shock absorbers and to the fuel feed - the TA/TB had a separate reserve and this was replaced by a low fuel warning light for the TC. The battery was moved, from behind the seats, to the more convenient engine compartment location. Abingdon later forgot how much more convenient it is to have the batteries accessible and moved them back behind the seats for the MGA and MGB!

In a world starved of cars, especially sports cars, the TC was an immediate success. It was exported to many countries where it was raced, rallied, used in trials and generally abused by enthusiastic owners. There had been quite a good sale for the products of the Abingdon factory overseas before the war but the need to "export or die" in the late 1940's saw a vast growth in these sales. At first deliveries to the United States were small but these had increased so much by 1948 that the factory produced a special version for that country with bumpers and turn indicators. However, as far as I know there were never any left-hand drive cars produced by the factory.

By 1949 it was obvious that the ageing TC would need replacement by something that was a little more modern. The Y-type saloon was already being produced at Abingdon and this was a much more up to date design featuring independent front suspension, originally designed by Issigonis before the war, allied to a stiffer chassis. Some urgent work resulted in the TD being built around a modified Y-type chassis, fitted with a twin carburettor version of the Y engine mated to a Y gearbox and YB back axle. This axle was of hypoid type and quite different to that of the TC. The front brakes now had twin leading shoes and the steering was rack and pinion. The car lost its large 19in. centre-lock wire wheels in favour of 15in. bolt-on disc wheels. It

A very early TD with plain wheels. This attractive factory publicity photograph shows the car off well.

was this last change that caught most attention. The new wheels were a lot more practical and "modern", and I think suit the styling of the car better than wire wheels, but a lot of people preferred sports cars to have wire wheels and these were not even an option on the TD.

Along with the mechanical changes, and the styling alterations forced by the adoption of smaller wheels, the TD acquired a wider body with more creature comforts and, most importantly, it could easily be built in either right or left hand drive form. It was in the export markets that most of the 30,000 sales of the new car were to be found. The wider track, better steering and slightly softer suspension made the car a lot more comfortable to ride in than the earlier cars and this was certainly appreciated abroad. There is no doubt that the TD sold well overseas and the considerable numbers that still survive have made it one of the cheapest of the traditional M.G.s to buy now. As I said earlier it is also a very nice car to use everyday and its affordable price means that it can make a practical alternative to some more recent sporting cars. There are a few modifications you can make to improve its practicality for modern roads and these we will examine later.

M.G.s have always been used in competition and the TD was to prove no exception. It had the disadvantage of extra weight, when compared with the TC, but the extra strength and roadholding were of help in making the car competitive. The XPAG engine was very easily tuned to give a considerable increase in power. The factory produced a mildly tuned competition version of the TD called the Mk ll. For this model the power was increased by using larger valves, a raised compression ratio and larger carburettors. To supplement the engine modifications the suspension was altered by the addition of friction shock absorbers mounted on the front suspension arms and on special brackets attached to the chassis at the rear. To avoid fuel starvation at high

A side view of the same TD. Note that the wheel rims have been painted a dark colour to tone with car.

engine speeds twin fuel pumps were fitted, with the fuel lines duplicated right back to the tank. The standard axle ratio was 4.875 : 1, as against the usual 5.125 : 1, although both the lower ratio and a higher 4.55 : 1 were available as options. After the first car, TD/C1123, was built this model was available throughout the production period but not heavily marketed as a separate version. Towards the end of production, from TD/C22613, the factory identified the Mk ll by separate badges on either side of the bonnet (hood) and on a plinth on the rear bumper. The badges on the radiator and spare wheel hub cap were now black and white, as fitted later to the TF. The cxistence of this Mk ll model, of which about 1700 are thought to have been produced, has caused confusion over the years. Many people have thought they had a Mk ll because the engine number quoted on the chassis plate started with the prefix "TD2". This prefix, however, refers only to the change from a seven inch to eight inch clutch and nothing else. The only identifying reference to the Mk ll appears in the chassis number. All Mk lls have the prefix "TD/C". Mark lls may be fitted with engines prefixed "TD, TD2 or TD3"; the TD3 engines were, however, only fitted to this model. The TD3 engine designation was introduced from XPAG/TD3/17029.

By 1953 the appearance of such new sports cars as the Triumph TR2 and the Austin-Healey 100 had made the TD look even more old fashioned. In an attempt to keep sales whilst a more modern M.G. was designed, the

The 1952 Motor Show and on the M.G. stand young film star, Simone Silva, tries out the driver's seat of an export TD. It is interesting to see the Wolseley 4/44, with a similar engine, in the background. Looking at the names of the other manufacturers it is hard to spot any still in business today.

team at Abingdon gave the old car a face lift. By lowering the height of the scuttle, incorporating headlamps into the wings, changing the interior and windscreen and fitting a more steeply raked radiator the whole appearance of the car was changed. Mechanically the new model, called the TF, was almost identical to the TD. However, the engine was uprated to Mark ll specification and the higher ratio back axle fitted.

The TF has now become regarded as one of the prettiest cars ever built at Abingdon. When it was new, however, it was seen as a poor substitute for a completely new car and received a very mixed reception. Obviously more power was needed to allow the model to compete with its faster contemporaries and the factory later fitted a 1,466cc version of the XPAG engine, called the XPEG, in an attempt to satisfy these demands.

Total sales of the TF between 1953 and 1955 were 9,600, of which around 3,500 were the TF 1500 model. The demand for wire wheels, which had not been available on the TD, prompted the factory to make them an optional extra on the TF. Somehow the styling of the later car seems to lend itself better to wire wheels than the TD and the option proved popular, especially in the United States. If imitation is the sincerest form of flattery then the TF has received a lot of this. Over the years many kit car manufacturers have produced TF look-alikes powered by everything from large V8 engines to rear engined VW Beetles. A serious attempt to recreate the TF, with modern mechanical components and up to date safety and electrical features, was made by Alistair Naylor in 1984 when he produced the Naylor TF 1700. Although well built and beautifully finished he was unable to obtain enough orders to keep the car in production and eventually the Company was taken over by Hutson who have kept on making them, in very small numbers, alongside their business

A TD Mark ll competes in a club event in the 1950's. The additional dampers fixed to the front supension arms of all Mark lls show up well in this picture.

producing spares for the original cars.

In the 1930s it was fashionable to build cars with aerodynamic styling and M.G.s offered a streamlined coupe called the Airline on both the P and N-type chassis. When production of the TA commenced this body style was offered but only one or two cars were built. A number of TAs were modified by their owners subsequently and at least one car was specially built by the factory with sixteen inch wheels and much heavier wings, more like the VA in style.

After the war the factory bodied three TC chassis as vans to use for works deliveries but these were not successful and were subsequently re-bodied as standard cars and sold. The registration numbers of these three cars were FJB 319, 320 and 321.

Quite a number of TDs were sold as chassis to coachbuilders. The most well known of these special bodied cars was the Arnolt TD produced by Bertone for sale in the United States through the Arnolt Corporation. Around a hundred cars were built of which two thirds were coupes. In addition to Bertone, Zagato, Castagna, Vignale and Monviso in Italy produced at least one car each and in Switzerland Reinbolt & Christie, Ghia/ Aigle, Langenthal and Worblaufen also built a few cars.

In America the Inskip Corporation, major M.G. dealers, produced TDs lengthened to provide four seats. The front and rear parts of the car remained unchanged but the wings were modified slightly and the doors lengthened.

In Germany Wendler built one roadster, one cabriolet and two different coupes, Schlomer built three roadsters and Weidenhausen a single car. The most interesting

The factory late model TC again. This time the car is at the B.B.C. Television studios at Alexandra Palace, showing a good view of the later hood with a single rear window. The size of the window is approx. 22"X 6".

Previous page: Two good factory pictures of the attractive TF. The open car has the optional luggage rack.

German production were the cars constructed by Hennfarth of Stuttgart and sold through German M.G. dealers as regular TDs. Apparently these cars were constructed without any drawings or patterns being sent from the Abingdon factory and as a consequence all the body parts are slightly different to the factory version. These cars were purchased mainly by American G.I.s, many of whom took them back to the United States. The different construction has caused problems for subsequent owners when they have tried to fit replacement TD parts to their cars. The shape of all the exterior panels differs slightly from standard, making the car over two inches wider. The wiring and electrical components mostly came from German sources and the interior trim and exterior colour choice were all different from the Abingdon product. Imagine starting to restore one of these cars now without realising its parentage. You would be returning parts to suppliers endlessly with complaints about poor fits!

CHANGES IN PRODUCTION.

TA.

Mechanical

From engine number MPJG/684 the gearbox was fitted with synchromesh on third and top gears.
TA 0652. Engine breather pipe modified.
TA 0824. Petrol tap and pipe modified.
MPJG/1018. Water pump fitted with oil cup in place of greaser.
MPJG/1139. Oil filter changed.
MPJG/1294. Clutch cover plate modified.
MPJG/1514. Oil filter changed again.
TA 1306. Hand brake cross shaft fitted with return spring to improve the "fly off" action.
MPJG/1605. Double valve springs replaced by triple.
TA 1877. Carburettor linkage modified. Slow running control was also modified from TA 1990 when it was fitted with a return spring.
TA 2232. A flexible section was added to oil pressure gauge line.
TA 2253. The chassis lubrication was modified and a "grouped nipple" system fitted.
TA 2254. Front spring modified.
MPJG/2622. Oil deflecting plate fitted to front of camshaft.
MPJG/2847. Valve spring cap modified.
MPJG/3053. Rocker shaft supports modified.
TA 2882. Steering column made adjustable and steering box ratio changed.

Other Changes.

TA 1250. At about this point the body style changed. The rear wings were widened and the petrol tank width reduced. The earlier cars are usually now referred to as "narrow wing cars".
TA 1770. Wheels changed from side laced to centre laced.
Sometime during the first year of production the carpet over the gearbox was replaced with a rubber cover.

TB.

Mechanical.

TB 0307. Crankshaft pulley modified.
XPAG/645. Tappet cover modified.

TC.

Mechanical.

From approximately XPAG/2020 to XPAG/2965 the engines were fitted with an attractive aluminium rocker cover, with a neat octagonal oil filler cap, in place of the simpler pressed steel one.
TC 2196. Service sheet 67 issued May 1947 shows the re-routing of the speedometer cable.
TC 3856. Service sheet issued December 1947. Hydraulic dampers added to carburettors.
TC 4251. Service sheet issued in September 1947. Tapered packing pieces added between front axle pads and the front springs to reduce castor angle. This modification was to reduce steering effort.
TC 5039. The steering box drop arm was re-designed - a bend was introduced in it.

Electrical.

Approximately TC 1850. The headlamps were changed from those with the "U" pattern glasses to the diamond pattern. The later glasses have a greater curvature. The headlamp rims also changed - they have more curve in section than the flatter early rims. These new lamps changed the appearance of the car quite a lot.
TC 3414. The control box was changed from Lucas RF91 to Lucas RF95/2. The wiring for the two boxes is identical and it is possible to use parts from the later unit to repair an early box. When the factory fitted the

The EX-U model. The only TC fitted with bumpers by the factory. Here we see front and rear views, note the M.G. badge mounted on a plinth. Special brackets mount the bumpers directly onto the chassis. The plates usually found behind the front dampers were removed.

smaller, later unit they drilled an additional hole in the bulkhead but did not stop drilling the now redundant hole.
TC 4739. The fog lamp was changed from Lucas FT27 to SFT462.

Other changes.

About half way through production (about TC 5000) the bulkhead was changed to one with shallower flutes. Around this time the bulkhead was painted the same colour as the body, rather than grey as previously. The engine colour was also changed around then from grey to dark red. Also at about this time the sill plates fitted in the door frame changed to plain aluminium from ones lettered "The M.G. Car Company".
Approximately TC 5178. In April 1948 the half tonneau cover supplied as original equipment changed from black to fawn.
TC 5380. Also in April 1948 the attractive walnut veneered dashboard was changed for one covered in rexine in the same colour as the trim. The centre panel was changed at the same time from black to a metallic tan colour with black lettering, although the fixing bolts continued to be painted black. The bases of the dash lamps changed to the same colour as the centre panel. However, I have seen a very original car some years ago with a tan panel and black lamps but these, of course, may have been replacements.

Fairly early in the production run there was a small change made to the hood frame. The side rail of the frame, just behind where it fitted to the screen rail, had a kink on the early cars and was straight later. About half way through the production run there was a further change; the mounting, pivoted bracket moved from the joint between the two rear hood bows to a little way along the front bow, which was now curved.

THE TC EX-U MODEL.

As previously mentioned there was a special version of the TC produced for the United States market to satisfy the demand for small British sports cars at that time. This model, carrying the chassis plate designation EX-U (Export - United States), had the following changes introduced to meet local needs:-

Full width bumpers with overriders were fitted fixed to the ends of the chassis. These were similar to, but not the same as, those fitted to the later TD. The centre of the rear bumper had the M.G. badge mounted on a plinth. The headlamps were changed to the smaller 7in. S.700 type lamps capable of taking the sealed beam units required in the U.S.A. The front fog lamp was not fitted and the switch on the instrument panel was replaced by a headlamp high beam warning light. The rear lamp was replaced by two high mounted Lucas 482-1 stop and tail lamps fixed to the sides of the petrol (gas) tank. To provide a number plate light a Lucas 467/1 lamp, similar to the TD, was fitted above the now high-mounted license plate.

Flashing direction indicators were fitted. The front side lights had twin-filament bulbs, to double as lights and direction indicators, whilst the rear brake light circuits were fitted with two Lucas LIR L3 brake switch override relays to interrupt the brake light if the brakes and indicators were used together. As the wiring diagram for the EX-U model does not appear in all versions of the manual, it is reproduced later in this book. The instrument panel was re-arranged to incorporate a Lucas SD84 L18 switch and warning light, mounted in the centre instead of the inspection lamp plug sockets. The other dials were moved, the ammeter and oil gauge to the centre, ignition/lighting switch to the left and horn/dip switch to the right. The panel was painted metallic tan but had no lettering. The anachronistic "Thirtilite" on the dashboard was replaced with a second map lamp.

The horn was removed from the front badge bar and replaced by two scuttle mounted Lucas WT614 horns, one high note and one low note, as on the TD; a relay was incorporated in the circuit. There was an interior mirror fitted centrally above the dashboard and the windscreen now had laminated glass to comply with U.S. regulations. The steering wheel of most, if not all cars, had a gold pearl finish rather than black. These tan coloured wheels are now something of a rarity.

OTHER EXPORT MODELS.

The TC was exported widely to all parts of the world and to meet local regulations some changes to specification were made. Unlike the special EX-U model detailed above these changes were not extensive and the cars did not have distinctive chassis numbering. All exported cars did, however, carry the "Made in England" plate mounted below the guarantee and patent plates. For the export models the 30 M.P.H. lamp was replaced with a map lamp, Lucas DF41, and the headlamps changed so that both of them dipped. For left-hand drive countries the fog light and horn changed places on the badge bar, even though the steering wheel stayed where it was - there were no official factory L.H.D. TCs. In some markets a Kilometre speedometer was fitted. Some cars were exported as "kits" for local assembly, most notably to Ireland. After the United States the most popular destination for exports was Australia - where there are still a lot of TCs.

TD.

Mechanical.

XPAG/TD/2985. Purolator canister type oil filter replaced with Wilmot Breeden oil filter.
TD 4251. One piece drum and hub introduced.
TD 6035. Grease retaining seal changed from felt washer to steel cap. This change was introduced because of problems with grease leaking from the hub and a service sheet and kit were issued to modify earlier cars.
XPAG/TD/6482. Water pump modified.
XPAG/TD/7576. Oil pick up moved to central position in sump.

A photograph of an ice skating star, Anne Rogers, visiting Abingdon and trying out a new TD. Anne's father, E.R.Hall, was a well known racing driver and gained much success racing M.G.s pre-war. Also in the photograph are, on the left, John Thornley, Anne's husband Eddie Ward [centre] and R.C.Jackson.

XPAG/TD/9008. Exhaust valve rocker arms modified.
XPAG/TD2/9408. In July 1951 the change of clutch to 8in. required new bell housing and modified clutch release fork. It is the change of engine prefix to TD2 that has prompted many to call all cars with this prefix "Mark IIs", when they are nothing of the sort.
XPAG/TD2/10900. Shorter dipstick and tube fitted.
TD 11111. Steering rack inner tie rod housing modified.
TD 12285. Threads on wheel studs and on propshaft and back axle changed to Unified. The studs and nuts on the spare wheel carrier were changed from car 12419.
XPAG/TD2/14224. Much improved oil pump with integral replaceable element oil filter introduced.
XPAG/TD2/14948. New finned sump in cast aluminium with increased capacity, up from 9 pints to 10.5 pints.
XPAG/TD2/15861. Improved clutch linings fitted.

The cockpit of an early TD. This looks to be one of the first hundred cars built as there is no evidence of the scuttle hoop being fitted. The trim panel in front of the door appears straight, if the hoop was there I would expect to see a bulge in the bottom of the panel to cover it. The dashboard exhibits the early TC style instruments and there is no water temperature gauge. You can see here the very distinctive grain pattern on the tan rexine used for the trim. Also visible are the black heads on the instrument panel fixing bolts and the raised heads of the chromed fixing screws for the dashboard. Later cars had number 10 Phillips head screws 1 1/4" long, but here this early car has screws with slotted heads.

XPAG/TD2/16482. Gearbox selector rails modified.
XPAG/TD2/17289. Shorter pushrods fitted with longer adjusting screws in the rocker arms.
XPAG/TD2/17969. New cylinder block with round water passage holes. Modified head gasket - also for use on all earlier cars.
XPAG/TD2/20942. Distributor fixing modified.
XPAG/TD2/20972. The oil pump modified to incorporate a priming plug.
XPAG/TD2/ 22717. From car TD 22251 clutch operating cable changed to the stronger rod system.
XPAG/TD2/22735. Cylinder head with round water passage holes. New head gasket and long reach plugs.
XPAG/TD2/24116. New camshaft fitted with revised valve timing. Tappet clearance altered from .19in. to .12in., noted also on new plate on rocker cover.
XPAG/TD2/24489. Oil pick up filter changed.
XPAG/TD2/26635. New oil pump body.
XPAG/TD2/27551. Improved specification material for crankshaft.
XPAG/TD2/27867. Head machining changed to reduce the height of the valve spring seats, leaving the valve stems protruding more.
XPAG/TD2/28167. Holes drilled to wire lock heads of bolts holding rocker shaft pedestals.

Electrical.

TD 7624. Headlamps changed on home market cars to left hand dipping.
TD 8142. Lucas control box RF95/2, as TC, replaced by Lucas RB106/1 and separate fuse box SF6.
TD 21303. Rear lamps replaced with circular type. Wings changed to suit.
TD 22315. Flashing direction indicators fitted to U.S. models. Wiper motor moved to centre of screen.

Other changes.

TD 0351. After only 100 cars produced, a hoop was added to the chassis to brace the scuttle.
TD 0501. The solid pressed wheels, as fitted to the Y-type, changed for ventilated type.
TD 4237. Listed as the point when a rubber mat was fitted to the driver's side on L.H.D. cars. At some time the pedals on L.H.D. cars were fitted with extension pieces and the floorboard on the driver's side gained a recess to give more room for large feet, so perhaps this was the point.
TD 10751 L.H.D., TD 10779 R.H.D. Speedometer and rev. counter changed from the chronometric instruments, as TC, to the magnetic type with dished dials. The ammeter also changed and a rheostat was fitted for the panel lights. From TD 13914 a combined oil pressure/water temperature gauge was fitted.
TD 17548. Headlamp high beam warning lamp fitted.
TD 18883. Foot operated dip switch fitted and horn button replaced combined horn/dip switch. Details were given to dealers enabling them to modify earlier cars.
TD 20374 L.H.D., TD 20696 R.H.D. Three bow hood frame fitted and sidescreen frames modified to suit. The frame was modified to improve headroom.
TD 20749. Additional mounting point was fitted each side of the chassis, alongside the scuttle hoop, for fixing the body.

TF.

Mechanical.

XPAG/TF/31263. Oil pump made self priming.
XPAG/TF/31943. The lower banjo coupling on the oil pipe carrying oil to the cylinder head had internal diameter reduced.
XPAG/TF/33024. Oil suction pipe and sump modified.
TF 3495. Dampers added to carburettors.
TF 3811. Improved front wheel grease retainers on cars with wire wheels.
TF 4760. Threads on tie rod ends changed to Unified.
XPEG/501, first TF 1500. TF 1500 badges fitted to sides of bonnet (hood). Red reflectors fitted to rear.
TF 6887. Wire wheels now had a deeper dished inner flange.
TF 6950. Last 1,250cc engined car.

Electrical.

TF 1501 XPAG/TF/31537. Fuel pump changed to high pressure SU HP-type AUA 57 mounted on a special bracket fitted to the rear of the chassis.
TF 8146. The casting for the Lucas horn changed (WT618).

CHAPTER TWO

BUYING AND RESTORING A T-SERIES M.G.

LOOKING FOR A CAR TO PURCHASE.

Looking around for a car to buy is always both a pleasure and a pain. Inevitably you will usually be told that the particular model you see is, at that time, "very sought after" or "the most difficult one to find"; this, of course, as a prelude to asking the highest possible price for the car offered for sale! Don't believe it! When looking for a T-series M.G. you have one tremendous advantage - there were a lot built and a high proportion still survive in one form or another. The large number advertised in all the motoring magazines each month means there is plenty of choice and you have no need to buy the first car you see.

Sensibly you should decide which model you would prefer. Mind you, that has not been the way I bought any of the cars I have owned! If you are not certain whether you would be better off with a TC or, say, a TF then maybe you should first try to have a ride or a drive in one or two cars. Perhaps just going to see a few without actually parting with the money is the most pleasurable part of looking for a car. If you are already in touch with other owners then there should be no problem. Most people are flattered to be asked to demonstrate their car and will be able to fill you in with full details of owning and running one like it.

Once you have narrowed down the field to a particular model then you should do your research before examining any suitable cars so that you can see if it is all that it is advertised to be. How original is it? Has it all the correct mechanical components? What about the small items such as instruments? I have tried to give as much detail as possible on originality in other chapters and hope this will be of help when looking at cars for sale.

Of course your choice may well be limited if you set a low budget for the purchase of the car. It is true to say that a full restoration is expensive and if you were to carry out a "no expense spared" renovation of any T-type at present you would probably be lucky to be able to recover your costs should you have to sell the car on completion of the work, even if you do not have to pay much for the original vehicle. However, as this is to be a hobby then perhaps it does not matter too much and maybe you would prefer to spread the cost by buying a cheaper car to start with and then paying out over a period to restore it. What I am really trying to say, not very well, is that it is possibly cheaper in the long run to look for a car that needs little or no work rather than a "basket case". That said, however, if, like me, you prefer to build up your own car then the cost will probably be secondary.

So, with your budget and model established, what do you look for when examining a car for sale? The first rule must be not to hurry any purchase. The worst thing to do is to buy a car in poor light or where you cannot get at it well enough to make a careful examination. Anyone wishing to sell you their car should be able to make it fully accessible for inspection. This is particularly important when examining a car that has been dismantled, as it is all too easy to buy what appears to be a complete car, only to find later that some of the parts hidden in all those boxes are wrong and you have to search for replacements.

Dashboard of TA with steering wheel removed. Note the price, £222, is just visible on the windscreen.

EXAMINING THE CAR

Go prepared! Take suitable clothing and perhaps some ramps if you intend getting underneath and having a good look. Also maybe a torch, screwdriver and, if you are looking for filler, a magnet. Also if the vendor has indicated that he is prepared to let you drive the car find out from your insurance company whether they will cover you. Most policies only cover for any damage you cause to other cars but will not cover the car you are driving unless this is first notified to them. Of course, you may be satisfied just to go for a ride, limiting your driving to trying the controls, clutch etc. in the driveway. Actually this may be all you need as you can learn quite a bit from the passenger seat if you ask the driver to, say, brake hard or accelerate up a hill. You will be able to get some feel of the car without worrying about driving it on, perhaps, unfamiliar roads.

When you first look at the car walk around it looking at the fit of all panels and for its general condition. If you are looking for a near perfect car and the one you have been offered does not come up to description perhaps now is the time to walk away from the deal. The more you look at the car, and the longer the vendor has to work on you, the more chance there is you will start to sell the car to yourself. I know it sounds a bit silly, but it is all too easy to start to gloss over faults by thinking "if I re-paint this, or have that re-covered.....". DON'T DO IT! If your first impression is that the car is not right, LEAVE IT THERE.

If you decide to examine it more closely, now is the time to have a careful look at the outside. Check that the car sits square, and standing away from it, that the front wheels, especially on a TA, TB or TC, look to be at the correct angle to the road. Look to see if the bonnet (hood) fits well. Does it line up both sides with the scuttle? Look at the wing (fender)/runningboard line. There should be a good flowing curve from the front of the car to the rear of the running board. A lot of replacement panels fitted over the years are the wrong shape and these result in the curve not being uniform - the only cure may be expensive replacement panels. Look down the side of the car. Are the front and rear wings parallel to each other? Are both sides the same? On a TC, for example, there were different front and rear wings fitted at various times and the car may have an ill matched set. Remember that early TA's had different rear wings and a larger petrol tank. On TDs the front wings were changed slightly when the front shock absorbers changed. If the TD is a Mark ll look to see it has the correct bonnet sides - there should be an extra bulge to clear the larger carburettors on all but the earliest cars.

At this stage, perhaps, you should look at the body tub. Unless you are buying the car with a view to replacing everything the condition of the body will determine just how much you will need to spend on renovation. Not only is a complete body very expensive but replacement will mean removing and re-fitting everything, apart from the main mechanical components fixed to the chassis. If you are going to do this you usually find yourself doing a complete re-build anyway as it is hard to ignore anything at this stage when you are going to fit a nice clean, new body to the car! Preferably when the vendor is looking the other way, take hold of the top of the body behind the doors and give it a good hard pull trying to move it from side to side. On a good car there will be little movement and you should just rock the car on its springs. On a bad one the whole lot could almost come away in your hands! Open and close the doors. Do they close and lock easily? Gently rock an open door to see if there is movement in the hinges and the door pillars. Look to see if the front door pillar is strong and the striker plate well fitted. Look at the fit at the front of the doors where they meet the scuttle; the top of the door and the flat area of the adjacent scuttle should line up. Originally the factory applied lead to this area to build up the scuttle to match the top of the doors. Crawl under each side and look at the main body side members and the metal supporting framework - both can deteriorate quite badly. On the TD and TF there is a lot of steel in the rear bulkhead area. This can rust and repair will probably mean replacing most, if not all, of the body. Finally look at the condition of the wings (fenders) themselves, look for signs of repairs, usually more easily seen underneath, and perhaps check for filler. The rear wings on the TC are screwed to the wood and on the TD and TF bolted to captive nuts fitted to the wheel-arches. Look to see if they are well fitted and that the wing piping looks right. An over-thick piping usually is present to disguise poor fit. Lastly look at the radiator surround. Does it look straight or are there signs of twists or repairs? New shells are available but they are expensive and I have yet to see one that looks exactly right. They are very difficult to make.

THE MECHANICALS.

In some ways the mechanical condition of the car is, perhaps, less important than having all the correct components there. That said, if you are paying the top price for a "good" car, then obviously all the parts should work well and not need anything more than minor tuning or adjustment. A lot of cars are advertised now as having "all numbers matching". What does this mean? I suppose it is meant to indicate that the car still has all the original parts as fitted when built. However, it may mean no more than it has engine and chassis number plates that match those that left Abingdon! It is very easy to fit replacement plates and have them stamped. However, I would concentrate on making sure the car is fitted with the right type of engine, gearbox and axle for the model and that all the ancillaries, carburettors etc., are correct.

If possible run the engine for a while and listen for any unwelcome noises. Remember that the valve gear on the XPAG engine is fairly noisy and this can be ignored unless excessive. Rev it up and look at the exhaust for blue smoke. After the engine has run for a while look for oil leaks especially from the bell housing. This is a weak point on XPAG engines and some oil is to be expected but if, especially after a run on the road, there

The back of a new TD chassis. This car is fitted with the earlier type of dampers and also has felt strips and mounting pads in place for the prop. shaft tunnel and the body.

is a large amount of oil dripping, you will probably need to have the engine out. Look for oil pressure, especially at idle - this will vary from car to car but you should see at least 20 lbs. at idle and over 40 lbs. at 2,000 r.p.m.. Do all the instruments work? Take off the radiator cap and look for signs of any oil in the water and examine the dipstick for any droplets of water in the oil. If the oil appears thick and frothy then water may be getting in somewhere.

Look carefully at the chassis. You can see quite easily from looking under the car from the front if the chassis rails are badly bent, but small amounts of distortion will be less easy to spot. Look for serious rusting, especially at the rear where there is not as much oil thrown around to protect things. The early chassis are less prone to hidden damage but the outriggers for the body can rust and I have seen some bad rusting around the rear spring hangers. The TD and TF are more difficult as the whole chassis is boxed and is of thinner gauge metal. I can only advise a good prod at hidden areas with a screwdriver and a general assessment of the amount of neglect of the car as a whole. In the end if the rust is not too bad it is unlikely that anything you find will not be repairable by one of the specialists in commercial vehicle chassis repairs.

A road test will show you how the car goes and if you have any experience from driving other similar cars you should be able to see how this one compares. Look particularly to see how an early car steers - they can be very good, but they can be awful. A TD or TF should have no lost motion at the steering wheel and should feel very precise.

TA REAR AXLE AND BATTERIES

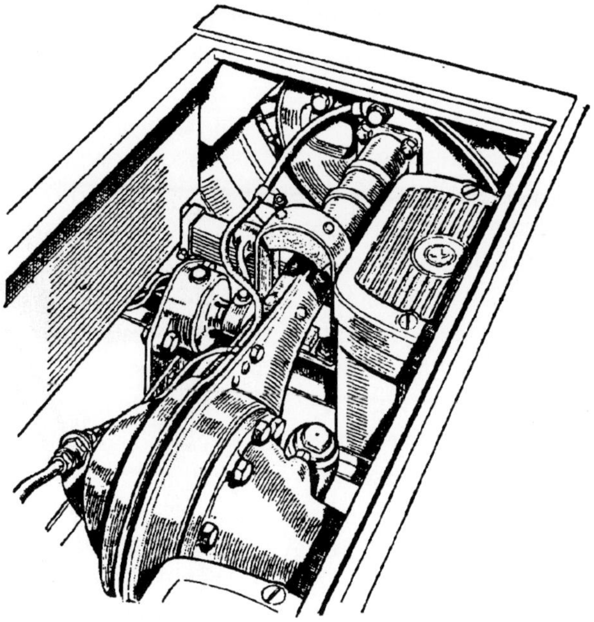

INTERIOR AND WEATHER EQUIPMENT.

You will probably have looked at this area when you first saw the car but perhaps a closer look is called for. Look carefully at the seats to see if they are leather and not plastic. Look at the colour to see if it is original (unless this does not worry you) and that the interior trim panels are correct. A complete re-trim is expensive. Look at the soft top, frame and side screens; these should be tan except for TAs, TBs and early TCs. A new set of weather equipment is very costly. Finally are all the seat runners in place and do they work. While you are in the car have a look at the instrument panel and see if the instruments and controls are as original. There are many reproduction units around and these may be fitted. You can sometimes see this by looking at the rear of the panel from underneath as the reproduction items often have bare, or painted aluminium cases rather than bakelite. This may not matter overmuch but you should know.

Well after all that examination do you want the car? If you are not sure go away and think about it. Should there be some work needed that you will want to entrust to a specialist, then ask one for some idea of the costs involved as this will have to be taken into account when deciding how much to pay for the car. If, however, you are happy GO FOR IT... you will be buying about the best fun to be had on four wheels and also joining a great number of other keen owners who will help you enjoy your car. M.G. people are nice people!

RESTORATION - GENERAL.

Before buying a car to restore, or starting to take a good car to pieces, think carefully whether you have the necessary skills, time, money and patience to carry the job through. Most restorations take at least four times as long, and cost three times as much, as your original estimate. Many of the "basket cases" started off as perfectly sound cars before being pulled apart by people with more enthusiasm than sense. One of the better known purveyors of early M.G.s claims to have made a good living over the years selling and re-selling the same cars to a succession of hopefuls, who never *do* manage to put the whole lot back into one piece!

If you do buy a complete car in working order you could, perhaps, consider doing a running rebuild, working on just one section at a time. For example, one year rebuild the engine and then have some fun with the car before starting, say, on the bodywork the next year. In this manner you can slowly improve the car without ever having it off the road for years on end. Of course, there are some of us who are never happy until we have completely stripped the car down to spend all our spare time for the next year or three in the garage. We are just as likely to sell the thing once finished and look for another heap to start pulling apart. There is a word somewhere for people like us.....!

WORK TO A PLAN

If you decide on a complete rebuild then some sort of plan can help. You will have an idea how much work you can do yourself and how much you intend to leave to specialists. Approach these early on to get an idea of costs, and of the time scale they work to - then double it! If the car is complete photograph every angle and talk to experts on the model to decide what is right and what is wrong with it. Take measurements and look at the fit of panels etc.. If the panels do not fit properly before you take it apart, try to find out why. Adjust the fit of the body, as described later, and line up panels so you can see how they should look. It is no good re-assembling a car you have taken apart and expect that the fits will just improve by themselves - they will probably be worse!

Take a good look at the way all pipes, wiring, and small fittings are mounted. MAKE NOTES AND DRAWINGS. You will not remember when the time comes to put it all back together. I speak from experience! I have found that it is a good idea to have a note book, or one of those wipeable boards, in the workshop so you can write down things as you see them. Before starting to strip down a very oily car it could perhaps benefit from steam cleaning.

When starting to dismantle the car it is a good idea to clean up parts as you go, even if you do not intend working on them right away. Sort out items for re-chroming and keep these separately. Use those sealable plastic bags to keep smaller parts in but do not forget to label them. Use a thick pencil, rather than a felt tip pen, on labels as pencilled notes do not fade away in strong light! Start a list of new parts you will need. These may have to be ordered early as some are only made in batches and you do not want to be held up in the final stages for the want of some small item temporarily out of stock.

Although it is not always a good idea to work to a deadline some sort of a target can help. Particularly if you decide to farm out, say, the engine rebuild to a specialist. Give him a realistic time to work to. Often he will be held up for parts and it is better to have the engine back rather earlier than you wanted, than after the deadline you need to meet to get the car to a particular event. Actually it is a good idea to work to finish some part of the car, like the instrument panel, at an early stage so you have something to show off to other members of the household to account for all the hours you are spending in the workshop!

Many restorers strip the car right down and then start first to restore the rolling chassis to show condition. As we shall see later, the chassis is the ideal jig on which to repair the body, or for assembling a new body, so perhaps the final finish should not be applied until after this work is complete. In fact, if only light repairs are needed to the body timbers, it may be better to do this before removing it from the car at all, because, if some of the joints are weak, it would be better to repair these before subjecting them to the stresses and strains of removal and storage.

WORKSHOP, TOOLS AND EQUIPMENT.

Many people have carried out superb restorations in their driveway, or by the side of the road, but for most of us some sort of dry, and, in winter, heated space is needed. Unfortunately a dismantled car needs about four times as much room as as it takes in one piece and, as most restorers have rather small garages, it is inevitable that storage of all these parts will be a problem. The two largest components of the car, the chassis and the body, are the most difficult to store. For this reason the longer they are kept as a unit the better.

Anyone needing parts for a re-build of a T-type can luckily call on the services of quite a number of specialists who are both knowledgeable and enthusiastic. Gerry Brown has owned, raced and re-built these cars for more years than he cares to remember and his advice is always available, both at Brown and Gammons and also at M.G.C.C. T Register rebuild seminars.

A sound body can, at a pinch, be stood on its back, but take care to leave the doors in place and brace it as best you can. The chassis can stand on its side against a wall and the lighter TC frame can even be hung on brackets. Smaller parts, such as panels, can be stored in the spare bedroom and the radiator shell would make a fine ornament in the sitting room!

The workshop or garage should have a strong bench with a vice (vise). The jaws of the vice should be at elbow height to be comfortable. The number of tools you have is governed by how much of the work you intend carrying out yourself, but there are some items that are essential. In addition to the usual hammers, screwdriver, files and so on, seen in most homes, you will need sets of open-ended and ring spanners with BSF/Whitworth jaws and also a socket set. The later cars also had some UNF bolts as these were being introduced to all British cars in the 1950s. Unfortunately you also need a few BA spanners as well as these threads are used on some of the smaller fittings. The BSF/Whitworth sets are becoming harder to find, especially outside the U.K., but most of the people supplying parts for the cars can help. As with all things, you only get what you pay for and some of the sets seen at autojumbles and flea markets are of poor quality. That said, it need not cost a fortune to set up your workshop and you can add to your inventory as you progress.

If you decide to carry out some specialist tasks, such as spraying or welding, then you will need the equipment. However, some can be hired and there are good but inexpensive sets on the market now - these will be discussed later.

ORIGINALITY IN RESTORATION.

Now the problems and arguments start! What exactly *is* meant by originality? A lot of cars are today described as "original" when what is meant is that they *look original.* An original car would be one that was exactly as built by the factory - never re-painted, rebuilt or had anything replaced. Well there are not many of those around now! Having cleared up that point let us look at what would now constitute an "original" car for concours judging.

Firstly, if it is to look as it did when it left the factory, then the use of a large number of unpainted chrome, or zinc-plated bolts on the chassis is not correct. There are a lot of photographs that show that the chassis left the works with all the bolts and fastenings painted black. On a number of cars I have taken apart it seems that they may even have painted chassis black over things like fuel lines, etc. after installation - however, I do not know if this was always the case. I do know, however, that the bolts were mainly just blackened steel, not plated, even when on the TC they were used to fix the

When re-assembling a car it is often difficult to remember exactly where every part fits. For this reason many photographs and good notes must be taken when the car is dismantled. Mike Beales, the lucky owner of these two TFs has the luxury of a complete car to refer to as the rebuild of his second car progresses.

bulkhead to the body. Most people now use some form of plating on these bolts.

Secondly it must be remembered that M.G.s were simple, inexpensive cars that were well built, but not finished to a particularly high standard, although all the fits of panels were very good. If you must, then by all means highly polish aluminium items, such as carburettors, but do not think that this looks original. There is also a tendency to have very high gloss finish on radiators and other black painted items but this is not how they left the factory. Actually I find that the satin black spray paints make the radiator header tanks look much better than do matt black or gloss. I have sometimes used this also to good effect on other engine compartment parts. The only trouble is that this glossy look on cars is becoming accepted as the norm and judges are marking more original cars down accordingly, as I have found to my cost.

I prefer to see cars in their original colour schemes. However, as with all parts of a restoration, the owner has the choice. You will have to live with it and if you like extra shine or a bright pink car with yellow seats then who am I to argue! Just don't ask me to buy it when you no longer want the car!

CHAPTER THREE

CHASSIS AND RUNNING GEAR

THE CHASSIS FRAME

The chassis of the earlier cars is really very different from that of the TD or TF, and of much lighter construction. However, the job of cleaning either is fairly tedious and time consuming, unless you resort to outside help in the form a sandblaster or chemical dipping. In my view if you can find a convenient way to save this chore you should use it. The only thing to remember is that you should cover the cleaned surface as soon as possible after sandblasting to avoid rust re-appearing. Before taking the chassis to be cleaned first try to see if there was any serious damage or corrosion that would need drastic work as you may wish to give the whole job to a specialist to do.

To check the frame you can first set up trestles or stands on the garage floor and ensure that they are packed until level, verifying with a spirit level. Set the chassis on these, and as a simple check, see if the chassis is lying equally on the stands. Whilst the chassis is set out, check all the diagonals against those in the manual. If you have reason to think your car has been in an accident then you must do this carefully. Sighting along the frame should reveal any gross errors and a certain amount of minor, or even major, tweaking can be attempted at home using strong pieces of timber as levers or car jacks as presses. For example, a bend inwards on one side member could be pushed out with a jack bearing on the other side as long as the load was well spread to avoid pushing the good side out. Do not heat the chassis unless you know what you are doing, as you may cause areas of stress that will cause later cracking. The diagrams below give some idea how minor errors may be corrected, but unless you have the necessary level surface to accurately check all the dimensions any serious damage would be better left to experts to remedy. The TA, TB and TC chassis are much easier to measure and correct.

Major areas of rust damage will have to be cut out and replaced. As I have said before you will be unlucky if the early chassis suffers much in this respect. The TD/TF chassis has more rust traps and being a box section, made of thinner steel, there is more chance you will have some work to do if the car has been used in a damp climate. If there appear to be areas of rust seriously weakening the metal then having the chassis shot blasted will reveal the extent of the damage. The rusted areas will have to be cut away and new steel welded in, but if the area is large, great care will have to be taken not to distort the chassis or lose the dimensions. If possible only part of the frame should be removed at any time so as to keep its strength. Work on the rest of the damage after replacing the metal removed. Unless the car has led a very hard life then the areas to be repaired will be small, and mainly confined to the rear, and body mounting outriggers.

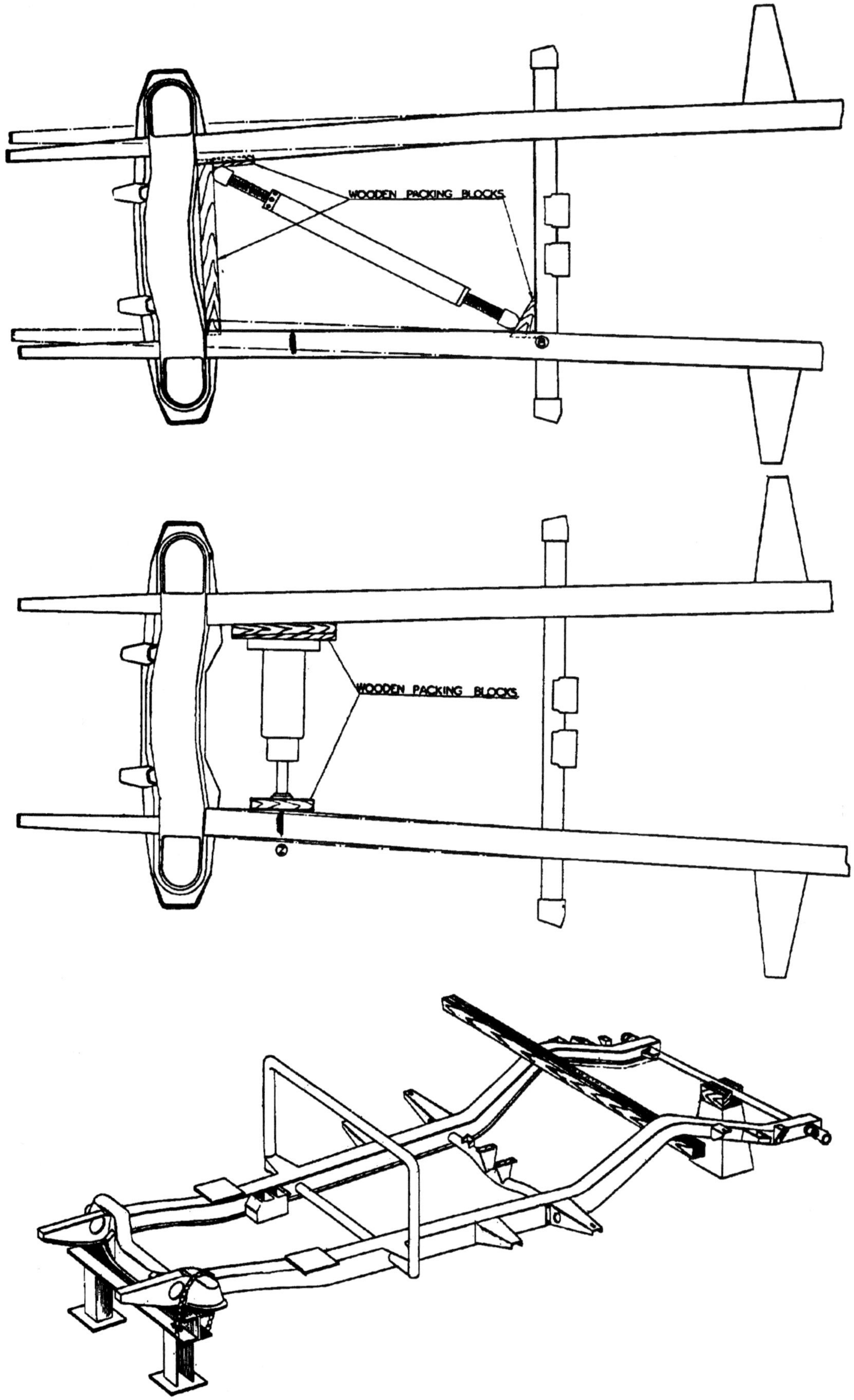

CORRECTING CHASSIS ALIGNMENT PROBLEMS WITH JACKS AND LEVERS.

The TC chassis sometimes cracks and to improve matters the later cars had a reinforcing plate welded in to extend the front chassis boxing rearward to almost enclose the handbrake cross shaft. It is very easy to see if your chassis has been modified and, if not, you should consider adding this extra strengthening.

The TA and TB chassis will also need to have the mountings for the spring rear trunnions examined. If the car has been run with these bronze trunnions worn then the steel chassis tube may have also been worn. To repair the rear chassis tube you can buy replacement end pieces turned from solid bar. The ends of these are stepped so they can slide inside the chassis tube after cutting off the worn part. The repair section is seam welded to the chassis. Remember to include the spacer between the end cap and the trunnions when re-assembling. The housing for the rear of the front springs usually suffers less from wear but is also less easily repaired. In extreme cases you will have to remove the rivets and either have the housing built up and re-bored to size or you could have it bored out to a larger size and have over-sized bushes made to fit. On both the front and rear springs there is a rubber dust cover that will need to be fitted on the spring before it is mounted on the chassis.

The finish given to the cleaned and repaired chassis is up to you. I usually use a good primer and then a good household enamel paint, but if you ever steam clean the car this will be quickly removed! A better modern solution would be black powder coating. This seems very wear resistant and looks good. I have seen good results from two-pack paints, but these, also, were applied professionally in a spray booth, so would not be suitable for home use. I am not sure there would be much of a cost saving over powder coating. I have used powder coating for smaller components, suspension arms etc., and the results are good. Some firms are equipped to sand or shot blast the parts first, and may even fill in blemishes between coats if a really good finish is asked for. I have also seen this process used, to good effect, on wire wheels. It seems the quality of the job depends on how good the company used is, so try the one you choose with some small items first.

SPRINGS

There is a temptation when restoring to replace more things with new than are needed. In the past I have unnecessarily replaced springs on a car because they looked tired and rusty, only to find that the new ones sagged quickly and were really no better than those replaced. If the car rode well on the old springs then it may only be necessary for these to be cleaned and repainted. The leaf springs on the TC should be flat under load and the shackles vertical. If they have sagged, usually the rear, then they will show a reverse curve under load and need attention. The rear leaf spring on the TD/TF, unlike the TC, still exhibits a camber under load and does not seem so prone to settle. If, however, you are not entirely happy with the old springs then all is not lost. It is possible, and not expensive, for these to be repaired by a specialist who can replace any broken or worn leaves and re-set and temper them to restore their original performance.

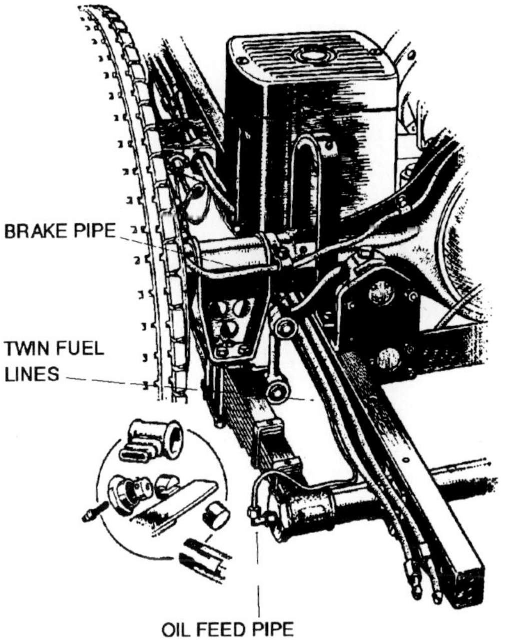

TA CHASSIS - DETAIL OF REAR SPRING TRUNNION.

TA/TB Springs. These are similar to the later TC and most of the following remarks apply to them but you will also need to examine the rear of the main leaf for wear. If the spring has been in use for some while, or the lubrication has been neglected, there will be wear on the leaf where it slides in the trunnions. If this is severe then either the main leaf or the whole spring will need replacement.

TC springs - front. The front ends of the springs are attached to the front of the chassis by hardened steel pins which fit directly into the spring eyes. This pin is drilled and has a grease nipple for lubrication. Usually just replacing this pin will take up the wear but if the spring eye is worn then a new top leaf will be needed. There is no interleaving in the springs so, after cleaning and painting, the leaves can be greased before assembly. The rear shackle rubbers should be replaced and, when they are assembled, they must be compressed so that they protrude equally from each side of the spring. The spring mounting plates usually only need a good clean but the special bolts that secure the front axle beam to the spring sometimes need replacing, particularly if the threads have been damaged when removing the locked securing nuts. These bolts are held to the top plate with countersunk screws that could be difficult to remove - the only solution may be a new plate as well, but this is simple to make. There was a tapered packing piece, thick end of the wedge to the rear, fitted by the factory after TC 4251 to reduce the castor angle. The idea of this was to make the steering lighter but I found that removing it from my car made the car less likely to dart about on bumps at higher speed; probably because of greater self-centring effect with the increased castor angle.

TC Springs - rear. As with the front these can be simply dismantled and cleaned. The front eye houses a replaceable "Silentbloc" bush so there will not be any wear on the spring eye. New bushes should be used along with new rubbers in the rear shackles. There is a diagram of the assembly of these in the manual but note that once again the rubber should protrude equally from each side. The large washer under the nut securing the lower link to the chassis is a special item the centre hole of which is shaped to fit a flat on the pin. This is to prevent it turning round and distorting the rubber as the nut is tightened. If you fit new or re-set rear springs do not be alarmed if the back axle is pulled down tight onto the bump stops on the chassis as it is fixed to the springs as when the car is laden there will be a little gap!

TD/TF Springs - rear. If your car still has the original springs then these had rubber pads at the ends of each leaf to stop the leaves chafing against each other. These pads are available and usually all you will need to do is to strip, clean and paint the leaves and then re-assemble with new rubbers. In addition to these there are also rubber pads on the spring clips that hold the leaves together and above and below the spring, between it and the mounting plates.

TD/TF Springs - front. Nice simple things, coil springs, all they usually want is a wash and brush up! If necessary they can be re-set but this is seldom needed. The assembly of the front suspension will be dealt with later under steering, all the details you need are in the manual.

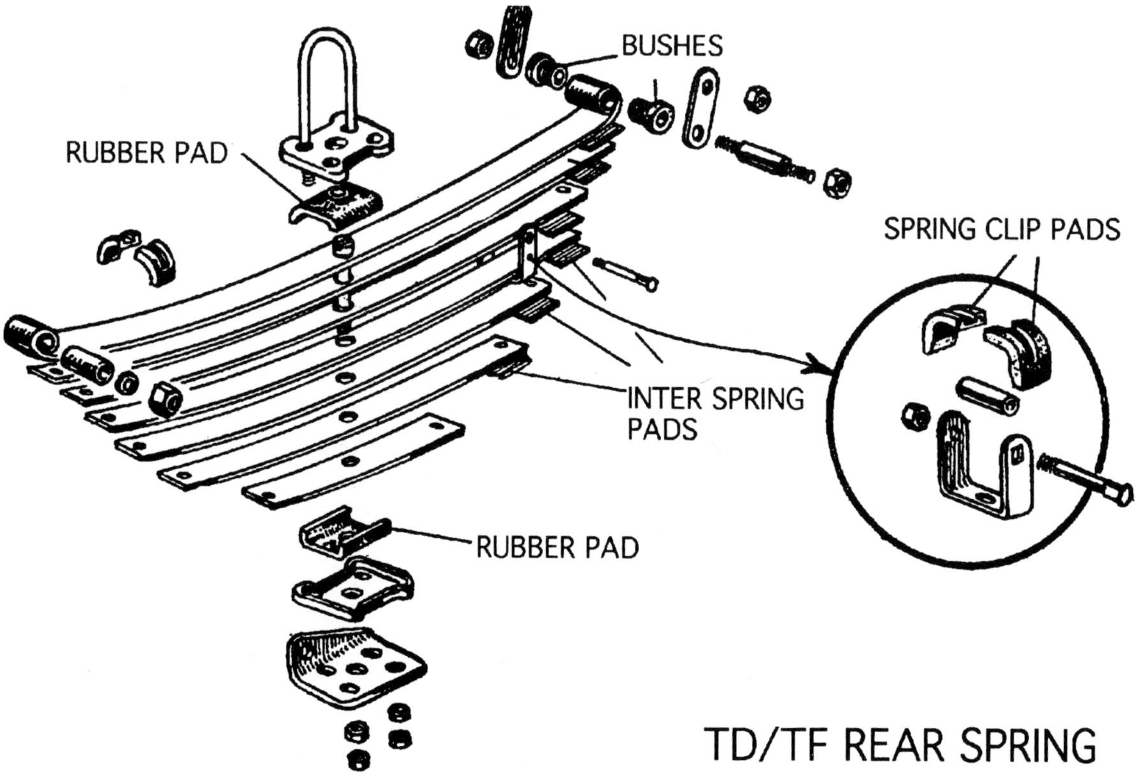

TD/TF REAR SPRING

TD/TF LEAF SPRING RUBBERS

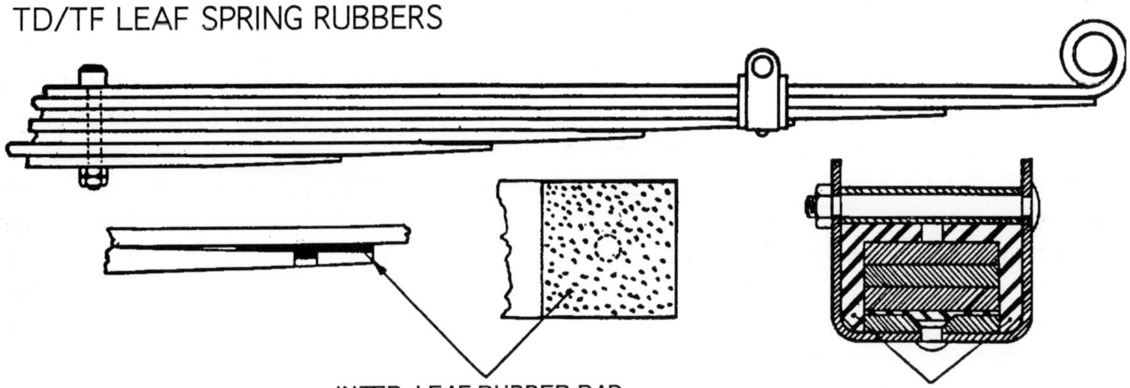

GROUPED NIPPLE LUBRICATION SYSTEM - TA/TB.

This system is a simplified version of that fitted to most of the O.H.C. M.G.s. There were three nipples each side and these served the front and rear spring trunnions and the handbrake cables. The main trouble with the system stems from owners applying grease rather than oil to these nipples. The grease clogs the pipes and the wearing surfaces are starved of oil. If the car is already fitted with this system the pipes should be disconnected and oil pumped through to ensure they are clear. If blocked it may be possible to try heat to remove the grease but it is probably better to replace the pipes. The tubing and fittings for the system are available.

SHOCK ABSORBERS (DAMPERS).

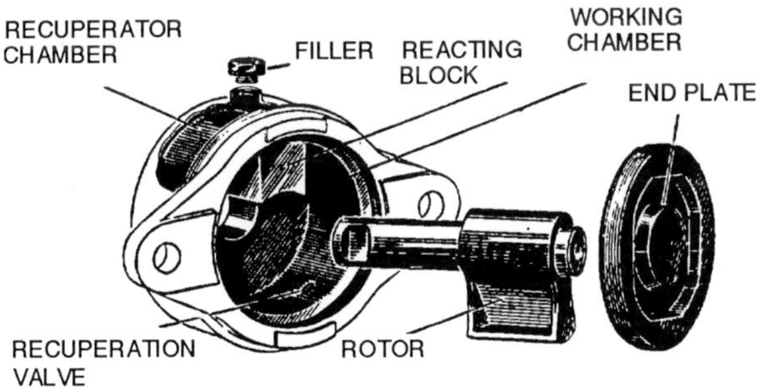

TA/TB LUVAX DAMPER

The front and rear shock absorbers on the TA/TB are the Luvax hydraulic type that act both on compression and rebound. They are adjustable by turning the regulator screw situated under the filler cap, although it is not advisable to tighten them up too much. Dampers of this type were never very effective and inevitably wear makes them even less so. One of the main faults seen, apart from the usual wear and fluid leakage, is being filled with the wrong type of damper oil. Using a thicker oil will *not* improve their performance and will accelerate wear, or even split the casing. As the clearance between the rotor and the case affects their performance, restoration is usually not a simple matter. At the time of writing some new shock absorbers are being produced to match the style of the originals. At the rear where they are not usually seen the TC type dampers could be fitted.

Both the TC and the TD/TF had hydraulic dampers, although the front dampers on the later cars also served the vital purpose of providing the top link for the front suspension. With all these dampers the problems fall into two areas, fluid leakage past the seals and wear in the rubber bushes in the link arms. The latter is easy to fix at home and the former more difficult for most of us.

The first thing to do, after external cleaning, is to ensure the damper is fully topped up and bolted back in place on the chassis or to a plate fixed in a vice. The arm should be moved over its full movement a few times to expel any air and then you should be able to see if it is working properly. There should be even damping in both directions with no "lost motion". If there has been a considerable fluid loss then it is probable that the seals are worn. Although it is possible to replace these yourself it is probably better to entrust the job to a specialist who will also have the necessary equipment to check the compression and rebound rates. Your only major worry is if the casing is cracked or the spindle so worn or corroded as to need building up and regrinding, or replacement.

FITTING NEW RUBBER BUSHES

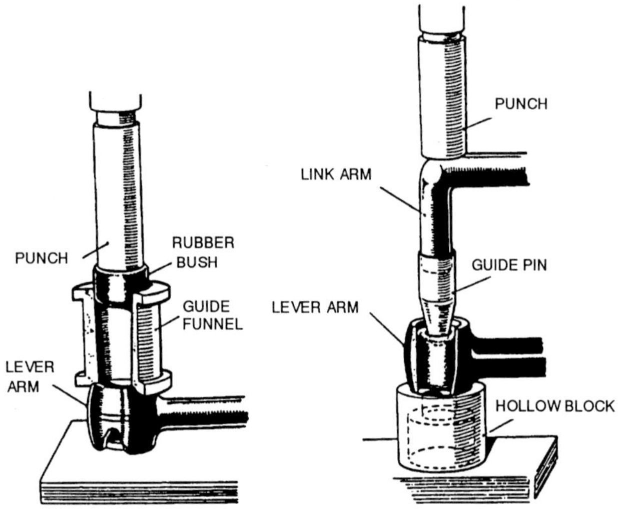

Worn rubbers in the links are easily replaced. If a pointed tool is made up to just fit over the end of the link arm this can be inserted into the rubber bush. The latest Moss catalogue lists a kit of tools to do this job. Using something like white spirit as a lubricant helps and this will completely evaporate later, not harming the rubber. Perhaps this is a good time for a reminder to make careful notes of how the links fit before taking them apart. It is a bit of a blow to have to remove the arm from a nice new bush!

ANDREX FRICTION DAMPERS FOR THE TD MARK II.

These dampers are really a development of the Hartford friction dampers fitted to many pre-war cars including all the M.G. Midgets prior to the TA. Contained within the casing are a series of fixed and moving plates interleaved with specially impregnated wood. These are under pressure from a plate spring and run in a bath of a heavy oil (S.A.E. 190).

The spring pressure is adjusted by turning the bolt, after slackening the locknut. There is approximately two and a half full turns of adjustment available with each quarter turn making a difference of 3 lbs. in. loading. The pressure is set by measuring with a spring balance at the end of the damper arm. The recommended setting is 24 lbs. for the front and 22 lbs at the rear. In practice for road use I suggest these are tightened until they can just be moved with a firm pull, which on a lever the length of the damper arm should be about right. The important thing is to have them all at around the same setting.

ANDREX ADJUSTABLE FRICTION DAMPER FITTED TO TD MARK II

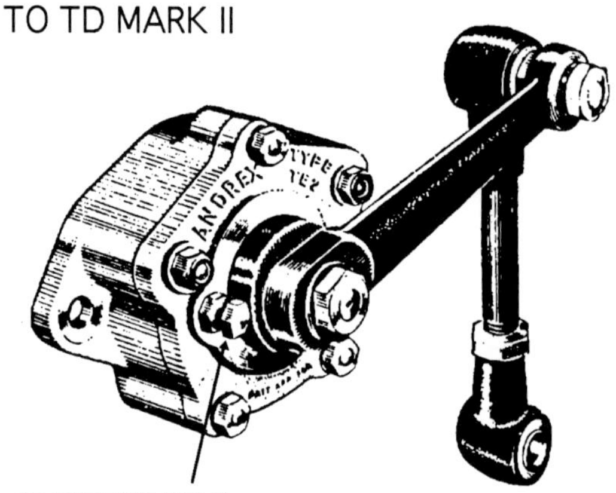

STEERING - TA, TB and TC.

It is probably true to say that there have been more rude words said about the steering box than any other part of the car! At best the steering box was not very good and although it works well enough, when all is properly set up, it does wear quickly and tends to show up any defects elsewhere. There have been many solutions offered over the years, especially from owners in the United States, and there is no doubt that some of these improved matters quite a lot. Some of the solutions centred around eliminating friction in all the joints and in the box itself, others concentrate just on the steering box. A recent and, I hear, most successful solution is to fit a steering box from a Datsun! The chairman of the New England M.G. T-Register, Dick Knudson, has fitted one to his much travelled TC and has been fighting off requests from other TC owners ever since to try it out!

Here, however, I will concentrate on points to look out for when working on a standard set up. The first and most important thing is to make sure that the front axle is not bent and that the eyes are not worn. If you place the axle on its pads on a flat table you can insert a 3/4 in. bar or dowel in the eye and measure the angle of this to the flat surface. The king pin inclination (the amount it leans in from the vertical) is 7 1/2 degrees and in the absence of any more sophisticated equipment you can check this with a simple protractor. Likewise the castor angle (twist backwards as viewed from the side) of 3 degrees can be measured. Although fairly crude you should be able to spot any major errors. Indeed, if you use long enough bars in the king pin eyes, then you will be able to see any large differences between each side visually. If the axle is bent a specialist should be asked to straighten it and then check the results. If you have access to a large press then you could try straightening it yourself but in any event do not use, or allow to be used, any heat. If the king pin eyes have been enlarged by wear then these could be shrunk with heat but only at the eyes and by a specialist.

TA, TB and TC Front Axle Assembly. With the axle correct take a look at the stub axles. The early cars had a stepped spindle and the later a tapered one. In both cases cracking and breakage are not uncommon. Have yours crack tested! They always crack at the inner end and this can sometimes be avoided by polishing this area to a good finish, to relieve surface stress. The stub axles have a left hand thread on the near side and a right hand thread on the off side. On both, the steering stop bolt is to the rear, to engage with the head of the king pin retaining cotter on full lock. If new king pins and bushes are fitted then the bushes will have to be reamed, after pressing them in. Make sure that the holes in the bushes line up with the grease nipple holes and that the groove to feed oil to the bearing face points the right way. This reaming is not as easy as it sounds as both bushes must be reamed in line. If you can beg, borrow or steal an expanding reamer with a mid size of 3/4 in. and a pilot, the job is easy. If the reamer does not have a pilot then it takes care and patience to ream both holes without removing too much metal. The king pin should *just* push through with hand pressure, Too loose and you start again with new bushes. The thrust washer fits below the axle. The wheel bearings should fit well on the stub axle. If the axle is a little worn try some "locktite". Do not forget the bearing spacer and the oil seal and ensure the dished washer is fitted with the dish outwards. In other words the centre of the washer is up against the bearing and the rim is away from it. New hubs are available and have right hand thread on near side and left hand on off (driver's) side.

TA, TB and TC Steering box. This will probably be worn if your car steered badly. Wear takes place on the top plate and the sector shaft. One of the main reasons for this wear is lack of lubrication. Originally there was a felt seal round the sector shaft/box joint but this was never very effective. The use of a too thin lubricant would not help. There is now a steering box lubricant sold which seems to about the right mixture of thick oil

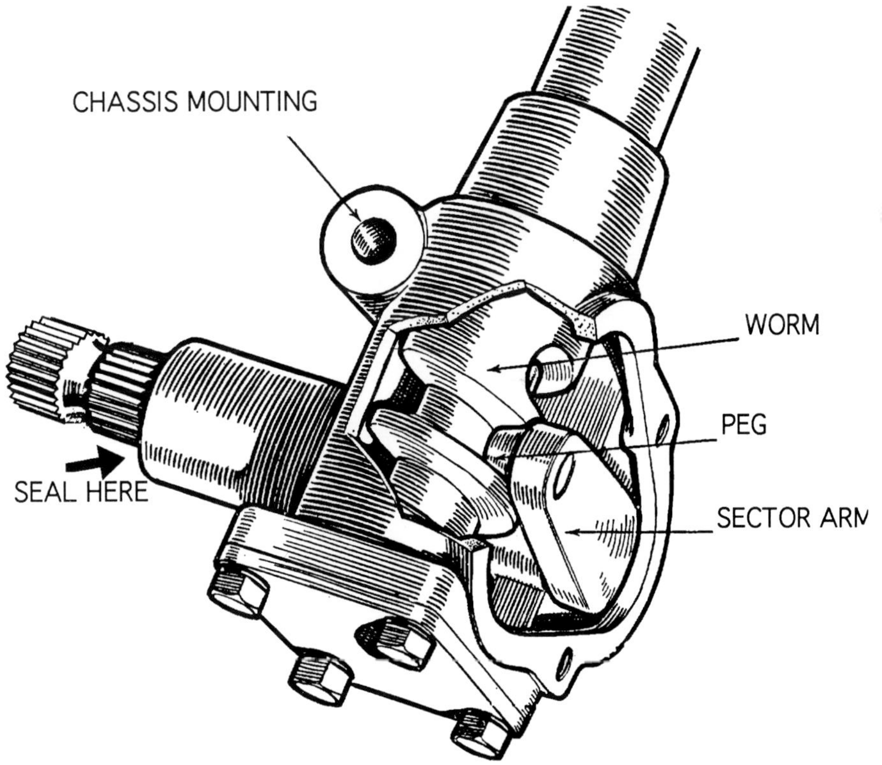

TA, TB & TC STEERING BOX

and grease to stay in place. Using a modern lip seal is also a must.

Looking now at the wear. The sector shaft will probably be worn where it passes through the box and the bore of the hole in the box will be elongated. The sector shaft can be built up and machined to size and a new peg fitted to engage in the worm. At the time of writing both new sector shafts and column and worm assemblies are listed as available by a major supplier. The sector shaft bore in the steering box will have to be bushed to return it to standard. This must be done with the centre concentric with the old hole and at right angles to the worm. The new bushes will need to be reamed in line to fit the shaft. The wear in the top cover can be removed by surface grinding. The shims under the cover are removed or replaced until the steering is *just* tightening up at the straight ahead position. The outer cover of the column should not be a loose fit in the box. If this has worked loose over the years then you will have to have this welded by someone able to weld steel to cast iron, or perhaps it could be brazed. There is a felt bush at the top of the steering column and this may need replacing. The new one should be very well lubricated.

TC Drop arm, track rod and drag link. The drop arm should be crack tested. If this fails there is no where else to go but the hedge! Likewise look carefully at the steering arms and the track rod and drag link ends. These are easily dismantled and new pins and springs are available with searching. Note that the pin that goes on the drop arm is different to the other three and the assembly of the associated joint differs. Full details are given in the manual, page 21. Make sure to split pin the nuts fixing the ball joints to the drop arm and steering arms.

TD/TF STEERING.

The alignment of the front cross member is vital to the steering geometry of the car. There are dimensions for checking this in the manual if you have reason to think your car has been in an accident. Unfortunately it is not easy to measure these accurately without setting the whole chassis up on a very flat surface; if the car is still in one piece the best idea is to have a full suspension check done at one of the tyre bays where they have a computerised suspension aligning rig. The cameras for the rig fit on the wheels with clamps so these will fit either disc or wire wheels. You will need to take the figures with you as the computer is unlikely to have them in its memory! Before dismantling check for wear by jacking the car up and rocking the wheel vertically to look for wear in the king pin trunnions, there should be very little movement. Also use a long bar to lift the suspension by the lower trunnion to try to detect any movement between the trunnions and the king pins.

Presuming all is well the assembly of the front suspension is simple and fully covered in the manual. All parts are readily available and the only word of caution is to make sure you fit the parts to the correct side of the car. The inner rubbers on the lower arms should be fitted when those arms are at their usual angle under load. These should not be fully tightened until the car is settled on its wheels. MGB V8 bushes will fit instead of the standard items and will last longer, as well as slightly improving the roadholding. If you are fitting the front springs to a bare chassis you will have trouble compressing them without special tools as you will just lift the chassis. You will either have to wait until the engine is installed or try to devise some *safe* way of wedging the chassis to the garage roof!

The steering rack is usually not prone to rapid wear. You should check the rack damper as shown in the manual and check and, if necessary, replace the tie rod ends and examine the inner ball joints on the tie rods. These are adjustable with shims. When re-assembling check the end float in the pinion (.002 in. - .005 in.) and the rack damper (.050 in. free play). If the rack itself is worn then replacements are available. In fact there are now complete steering assemblies available to service the demand for racks to convert from left- hand to right-hand drive.

The rubber bearings situated half way up the steering column usually need renewing as they rot and deteriorate, leading to odd rattles and knocks from the steering which are sometimes hard to trace.

TA, TB and TC REAR HUBS.

Oil leaking on to the rear brakes from the axle has long been a source of trouble for owners. The main reasons seem to be oil working its way past the seal in the bearing carrier and on to the linings. It also leaks down the splines of the half shaft, where it is pressed into the hub, and runs down the wheel spokes.

To stop the first leak it is necessary to make sure that the

TD/TF STEERING

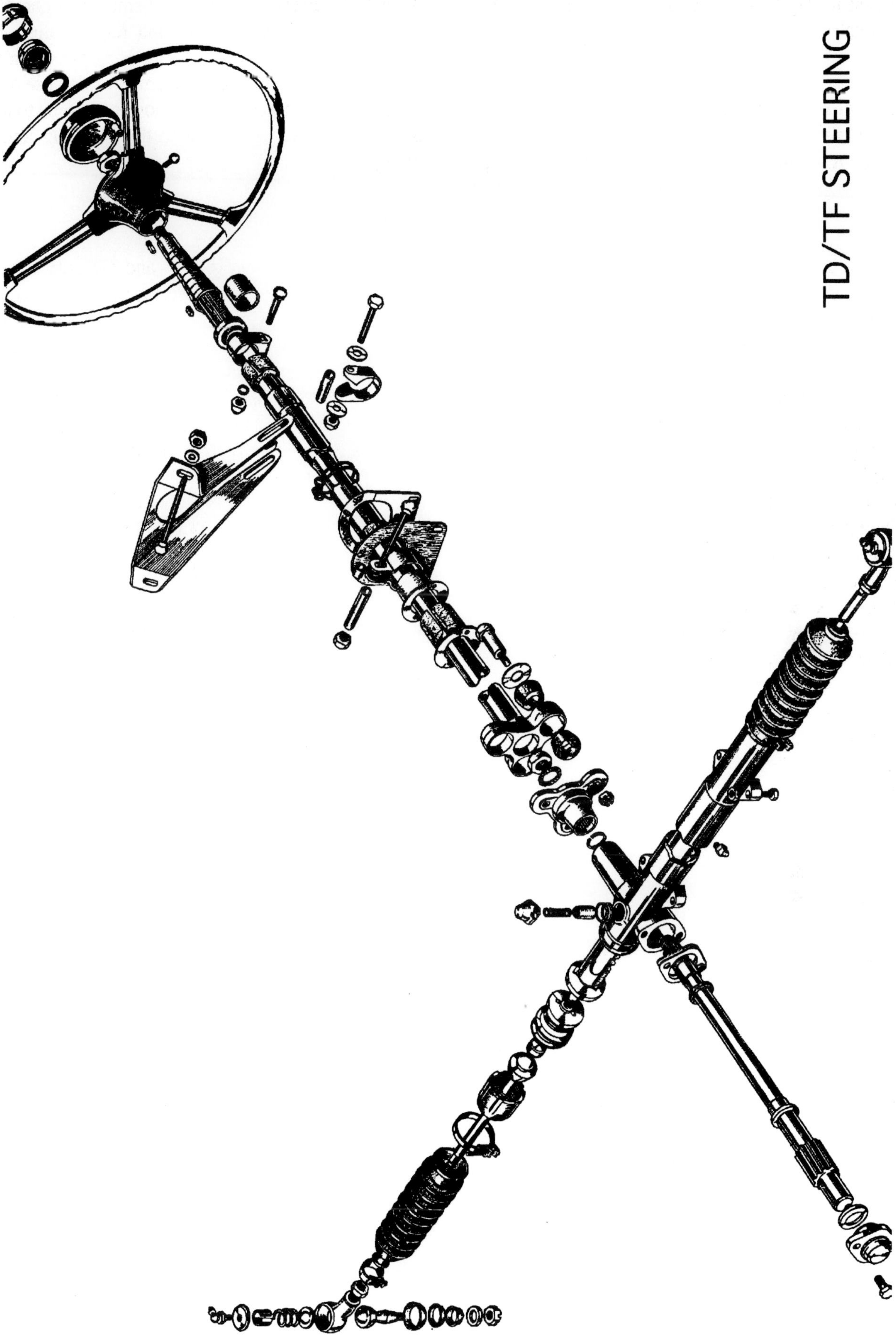

A restored TD chassis. As an early car it is fitted with the Luvax Girling dampers, replaced on later TDs and on TFs with the Armstrong dampers seen below.

half shaft is a good press fit in its hub. Wear in the splines of the shaft or hub will mean the hub twisting under cornering loads. When the hub is clamped to the bearing carrier the lip on the inner edge should press against the bearing to clamp it in place. The usual fault on assembly is that too thick a gasket is used between the hub and the carrier allowing the bearing to move. Before installing the gasket, clamp up the joint and measure the gap with a feeler gauge. If no gap exists then you will have to make up a shim to go between the hub and the bearing to give a sufficient gap. Service sheet 60, reproduced here, gives the idea. Another common fault is to use the wrong holes in the hub to insert the two fixing screws. The screws go through the *untapped* holes in the hub. Screwing through the tapped holes means the two surfaces cannot be pulled tightly together. I use a silicone sealer for this joint rather than a gasket.

The leak down the half shaft is best cured by cutting a disc of card the size of the end of the hub. Then, after wiping the whole area clean with solvent, put a generous amount of silicone sealer down the bore and push the cardboard disc down after it to form an oil tight seal. I know there was a grease nipple fitted in here but its use escapes me - there is so much oil getting to the bearing anyway, further lubricant seems pointless.

Service Information Sheet No. 60

METHOD OF CLAMPING REAR HUB BALL RACE

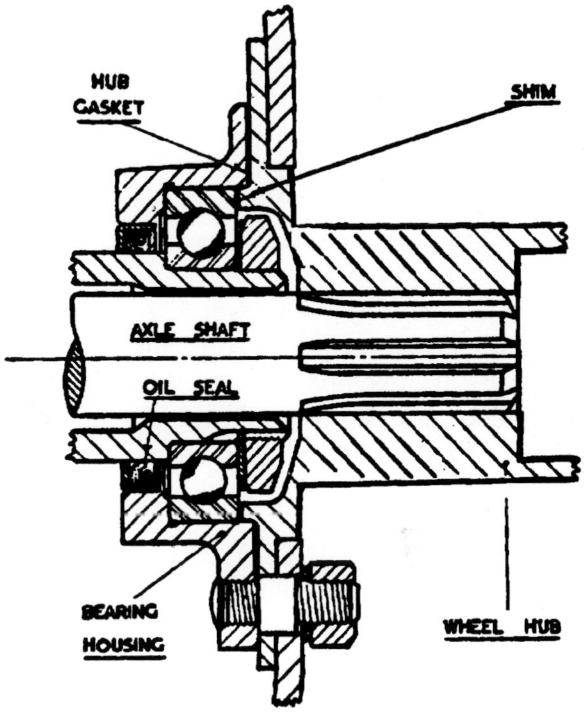

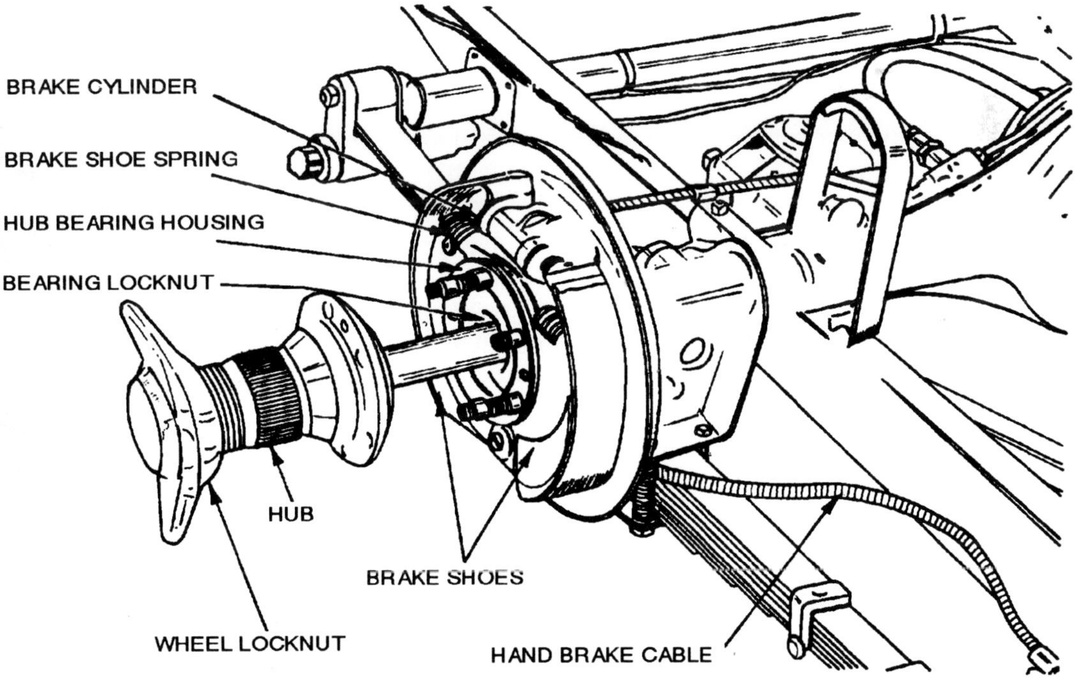

TC REAR AXLE AND SUSPENSION

PEDALS AND STEERING CONVERSIONS.

As part of the chassis work the brake and clutch pedal bushes and shaft should be examined for wear. The bushes are easily pressed out and replaced although the new bushes may need reaming to size after installation. The draught excluding boot should be slid over the pedal arms before they are fitted to the shaft and on the TD/TF before the pedals themselves are fitted to the shafts. The pedal rubbers are best glued to the pedals to avoid the chance of them coming off in use.

Quite a number of TDs and TFs have been repatriated to the United Kingdom from overseas in recent years and as most of these are left hand drive models the question is often asked "just how easy is it to convert these cars to right hand drive?" Later M.G.s, the MGA and Midget for example, were designed from the outset to be built with the steering wheel on either side. The TD and TF had different chassis for left and right hand drive models so the conversion is not quite so straightforward. If it is done as part of a complete rebuild then access to the chassis for the conversion will be easy but otherwise quite a bit of dismantling is needed.

Firstly let us look at a list of the major components needed for the conversion:-
STEERING RACK. The rack was only used on T-types and is not easy to find. The Y-type rack is similar and the casing can be used but the pinion shaft is shorter than on the T-type so a new shaft and rack gear will be needed. There have been a number of completely new racks produced recently as well as conversions of left hand drive racks but I have not seen either of these myself. At the time of writing the cost of producing a new rack using some of the parts of the old rack, but with new gears and casing, is just over £400. To this cost must be added about £250 for the other parts required to convert the car.
PEDAL ARMS. The pedal arms on the L.H.D. cars are of a completely different shape to the R.H.D. type and the shaft design also differs considerably. The clutch pedal is keyed to the shaft so that its movement can be transferred across to a lever, also attached to the shaft, which operates the clutch. On the R.H.D. cars the clutch cable, or rod on later cars, works directly off the pedal. For conversion, new clutch and brake pedals and new shafts are available.
PEDAL BOX AND SIDE COVER, PEDAL SHAFT BUSH AND MASTER CYLINDER MOUNTING PLATE. None of these items can be re-used but luckily are available new.
THROTTLE PEDAL AND ARM. Available new complete with mounting blocks.
TOE BOARD. The old item could be cut to fit but as it is a simple item to make, or have made, it looks better if a new one is fitted.
DASHBOARD. On a TD this will have to be replaced along with the board behind it. These are both available new, or could be cut from plywood.
BRAKE PIPE. The pipe from the master cylinder to the three way union will have to be replaced.

The car will have to be stripped down sufficiently to gain easy access to the pedal box area on both sides of the car. At the very least this will mean removing the gearbox cover, seats and floorboards as well as the toe board. The car will also have to be raised high enough to be able to gain safe access underneath. Removing the rack and steering column can be done quite easily without removing too many other components and the procedure is in the manual. Once the pedals and shaft are removed and the master cylinder unbolted, the old pedal box and mounting plate can be cut off with an angle grinder. If this is done carefully and the old hole for the pedal shaft bush welded up, all signs of the left hand pedal mounting will be removed.

The new box and mounting plate are offered up to the right hand chassis rail and the position of the hole for the pedal bush marked. The chassis will then have to be drilled to take the bush. It is important for the position of the hole to be right as it must place the pedals correctly so that they fit in the pedal box without fouling the bottom of it. The brake pedal must line up with the master cylinder and the clutch with the clutch cable or rod, which passes through a hole in the front of the pedal box. Some adjustment of their relative positions will probably be necessary before finally drilling the hole.

With the bush just pushed into the hole the pedals and box are trial assembled and if all is well the bush, box and mounting plate can be welded in place. It seems hardly necessary to mention here that the welding must be well done - a failure could have fatal results! The bush may distort slightly when welded and a light ream with a suitable reamer may be needed for the shaft to fit. The pedals can be assembled and tried out to make sure that neither they, nor the operating rods, foul the pedal box. The master cylinder and the new length of brake

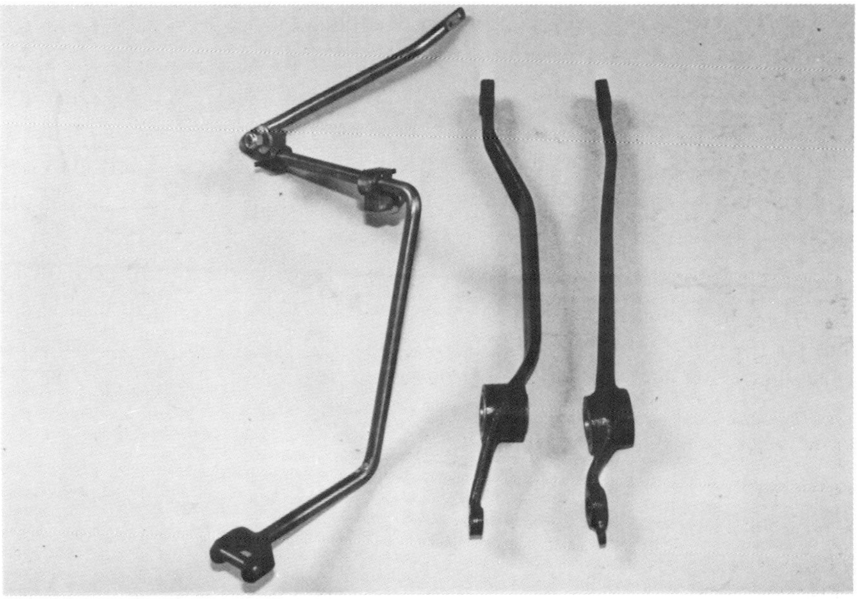

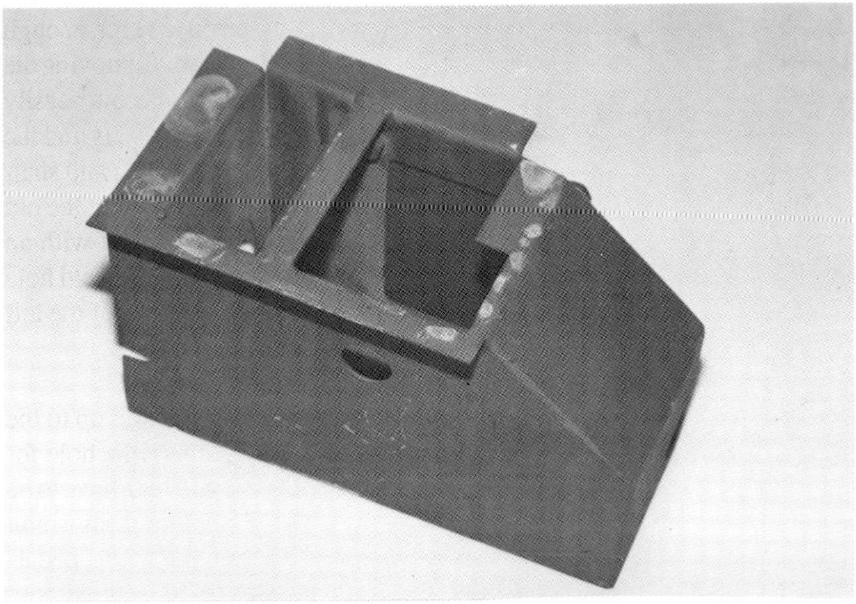

Conversion of L.H.D. cars to R.H.D. Top picture shows the pedals, all of which differ from those used on L.H.D. cars. The second picture is of the pedal box. The lower picture illustrates the mastercylinder mounting plate and the tube that is welded into holes in the chassis.

pipe are then fitted and the brakes bled. The replacement rack in place the steering column can be fitted and the new toe board bolted down. The throttle pedal is then installed and its shape adjusted to fit if necessary. The dashboard on the TD will have to be covered and fitted although the wiring and cables should swap over without much difficulty.

On late TDs and all TFs the dip switch will have to be transferred to just below the battery box. The mounting plate for R.H.D. cars was different but the old plate can be bent to fit its new position. The floor board on the left hand side of later TDs and all TFs had a cut out for a box that gave more room for the driver's feet. A new floor board will be needed as well as a new carpet. There will need to be some changes to the floor board and carpet on the right hand side as well. The pedal extensions on the pedals of the later L.H.D. cars can be omitted on R.H.D. cars unless they are needed for shorter drivers. Do not forget to change over the wipers on earlier TDs.

CHAPTER FOUR

BODYWORK

As we saw when we considered purchasing a T-series car the condition of the bodywork is very important when deciding which car to buy and how much to pay for it. The cost of restoring the bodywork on any neglected car is likely to well exceed the price of a full engine re-build and it will also take a lot more time unless a lot of the work is left to others.

Even with cars that have had a sheltered life in a dry climate there is almost bound to be some work needed on any car still wearing the original body. On a lot of cars, as soon as you start to pull the thing apart, it soon becomes clear that almost every part of the body will need repair or replacement and it is at this time despair sets in. Many a car has progressed to the stage of a fine rolling chassis only for the restorer to give up hope of competing the job when examining the pile of crumbling woodwork and rusting metal that had once looked a fairly sound body when removed from the chassis a year or so earlier. The moral of this is to look at the body and consider how you are going to set about it before embarking on the rest of the restoration.

Restoring a car involves many skills and few are lucky enough to be able to feel entirely at home with all of them. Some of us are very good with mechanical components and have no difficulty appreciating the fine engineering tolerances needed when working on the engine but are completely thrown by the idea of wiring the car or trimming the seats. However, the one part of the restoration most feel will cause no problem is the timber frame. Memories of school woodwork classes and DIY efforts around the home give us a familiarity with the tools and methods and it all seems fairly straightforward. Well they are WRONG! In spite of its slab-sided appearance the body of the T-series M.G. has almost no straight lines in it. There are no detailed drawings available and with most pieces of timber in the body having curves in two directions, and being mounted to their fellows at complex angles, the chance of errors creeping in are many.

At an early stage you must decide if you are to repair or replace. It is very nice to be able to keep as much of the original car as possible and I have always tried to do this wherever I can. However, it is certainly not easy to replace just a few parts of a body without causing trouble with fits of doors etc. unless great care is taken. Perhaps before this decision is taken you should look at your body tub to try to see how good (or bad!) it is.

The first step must be to establish just how well the body fits on the chassis and it should be made sure that the car is standing absolutely level and square before looking at it. The doors and bonnet (hood) are the crucial parts. It is the fit of these that first meets the eye and you will usually get away with other small errors if these areas look right. The top of the doors should line

A new replacement TD body fitted with doors and scuttle. This is the easiest way to tackle a re-build of a T-type!

up with the scuttle and should fit well against the body. The fit of the door can be adjusted by packing with washers between the body and the chassis, or removing packing already fitted. Before fitting packing make sure that the door is tight on its hinges. Also look to see that the bolts holding the body are tight before checking the fit, although these will have to be slackened off before packing can be fitted. If the front of the door is high this can be lowered by tilting the rear door pillar forwards by packing up the rear body mounts but this closes the door gap and may cause it to touch the front pillar. The alternative is to remove any packing under the centre mounting, or pack up the front, after slackening off the scuttle mounting bolts. If the door is low, removing packing from the front or rear, or inserting it under the centre mounting, will help. Quite a lot of adjustment can be found this way and fits can be improved vastly. Note that the factory recommended the use of felt or cork pads as insulation between the body irons and the chassis. This seems to break up quite quickly so perhaps a stronger material is called for. I have used thin stiff rubber or even a piece of vinyl floor tile in the past.

The fit of the bonnet is not so easily corrected. Each bonnet was adjusted on the assembly line to fit the car for which it was intended. I am assured, but have been unable to confirm, that there was a man employed in the TC days at Abingdon, with a large pair of tin snips, whose job was to trim the edges of *painted* panels to make them fit! If true it would explain the differences between panels from different cars as no two are alike. I tried recently to fit bonnet tops from one TC to the sides of another but they were miles different. The radiator can be adjusted with the stay rods and the mounting bolts packed to raise it, after slackening off the hose clips. All this will have a small effect but an over-large gap all round can only be reduced by moving the body forward. This will also affect the fit of the wings to the body but some movement here can be tolerated. The gap can also be adjusted a little by re-positioning the bonnet rest rubber in relation to the edge of the body cladding.

REPAIRING THE BODY FRAME.

If it is decided that minor repairs are all that is required, then these are best done before removing the body from the chassis frame. In most cases the damage will be in the door pillars and the body frame below the door. Assuming the car is now stripped down, with the bonnet and wings off, you will have to remove some, if not all, of the metal cladding. If this is in good condition take care as you will want to re-use it. In any event keep all the panels as patterns.

The cladding on all the T-series cars was steel, pinned to the frame with steel pins. The ends of the scuttle above the doors were bent over the top of the turned in edges of the side panels and the whole area was filled with lead to blend it into the line of the top of the doors. When removing this identify where the pins are, and remove them. This may not be easy under all that lead but I have found that if you free off the front of the scuttle panel, by pulling out the pins, and raise it up enough you can usually reach the pins by the doors from underneath and cut them off. If the panel is removed without disturbing the leading too much this will help later. If, however, the scuttle top is being fitted on a new body then heating the joint to remove the lead and extracting the pins before separating the top and side panels will not matter. The whole area will have to be re-leaded to make it fit the top of the doors on the new body.

The most vulnerable panels to rust are the side panels. On the TA, TB and TC this panel extends from below the scuttle to the line of the lower hinge on the rear door pillar. On the TD/TF the panel breaks just below the rear of the door. This means that the fortunate owner of one of the earlier cars will be able to remove most of the rust with the replacement of the one panel, whereas restorers of the later cars are usually faced with repairs to part of the rear panel as well. Replacement panels are sold but these may not always fit your repaired wood all that well and it may be better to make, or have made, a panel to fit your car exactly. We will look at making a new side panel later.

FITTING IN NEW TIMBERS.

All the timber used on the frame was ash. This did not seem to have been given much protection when new and some rot is bound to have taken hold. In dry climates this may not be severe and the metal may still be intact. There is no doubt that the frames of the TA, TB and TC are much worse than the later models. Not only do they contain more timber, but the more flexible chassis gives them a harder time. Ash can be obtained from timber specialists and cut to size using the old piece, or what is left of it, as a pattern. This really calls for a band saw. The first cuts from a flat piece of wood are not hard to judge but when cutting the curves in the other direction this is a lot trickier without a flat surface to work from. The timbers around the bonnet top and sides are easy as they only curve in one direction but the door pillars are very difficult. You may find it a lot easier to buy replacement timber parts from one of the suppliers. As with the metal panels they are seldom exactly the same as the originals but with wood it is

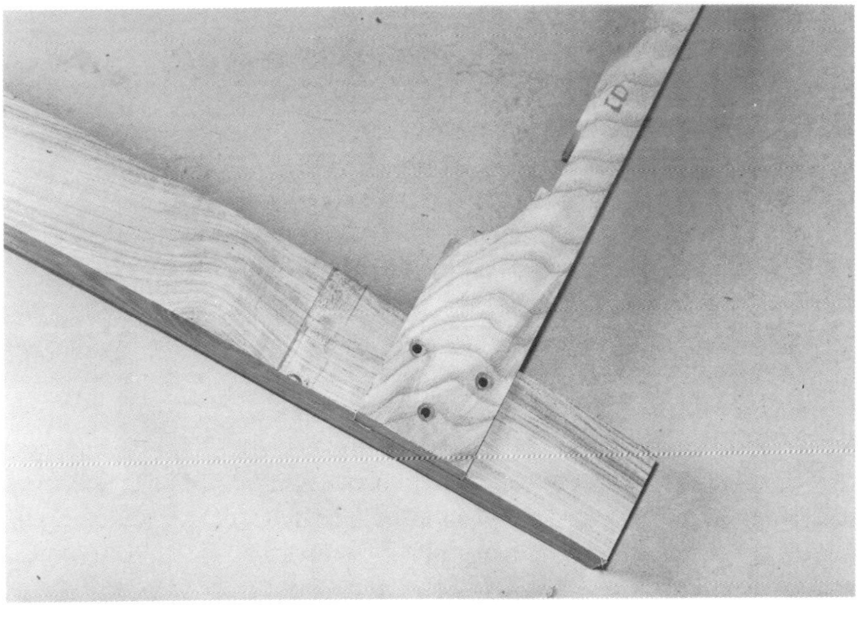

The rear door pillar/ main side frame joint on a TD. The recess just in front of the joint is for a small piece of steel to be let into the wood under the join in the panels.

easier to adjust the fits. This is not meant as a criticism of these parts as I well appreciate the difficulties involved and the differences between cars. In fact if you were to use a complete set of parts to build a new body they will usually fit together well, but using one or two bits on an old body can cause problems.

It is very important when working on the body not to loose the overall dimensions as the timbers are replaced. For this reason keep a careful note of the sizes of the door openings, height above chassis of scuttle and so on. If possible work only on one side at a time. If the front and back door pillars and the lower timbers are to be replaced, remember the rear door pillar is vertical as viewed from the side. If the rear pillar is fixed first and the door replaced, use the door as a template to ensure that the front pillar is in the right place, and that the door fits. Remember to place Tee nuts, or tapped plate, in the front door pillar for the door lock, and check the position of these and the locks before skinning.

A topic that will often raise its head when discussing body frames, is whether or not to glue joints. As far I have been able to see there was no glue used when the bodies were originally built. Accepted wisdom says that the wood could split when the chassis flexes if the joints are too tightly bonded. I am not sure if this is the case and anyway I have glued some joints, especially those in the timbers over the rear wheel-arches. The choice is yours but if you do glue anything use a waterproof exterior grade glue. The wood screws were originally plain steel but you could use stainless if you

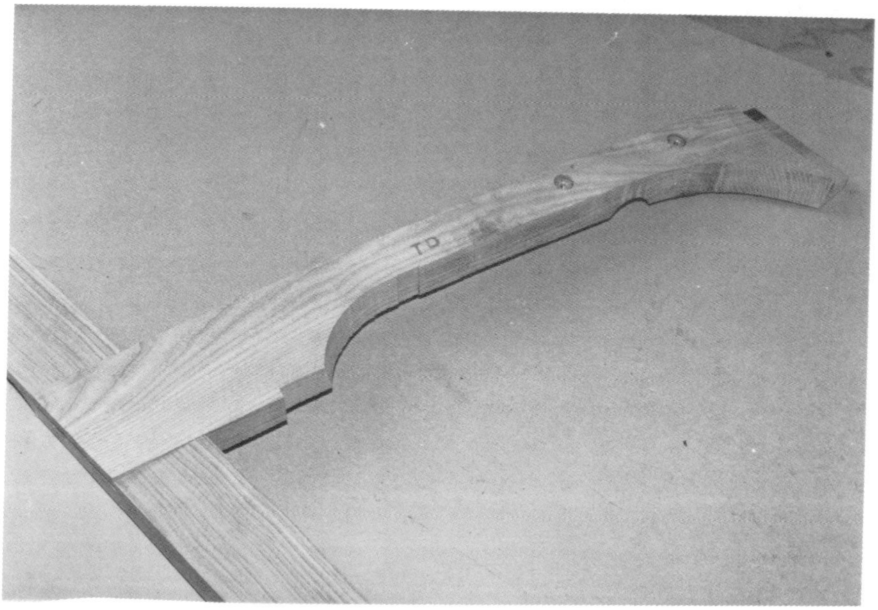

Front door pillar on a TD. There are two Tee nuts let into the pillar for two of the door striker plate bolts. The position of these should be checked before the body is skinned.

A TF rear body tie bar. The TD one is similar and both can suffer from rusting where the cars have been used in damp climates. The TA/TC bodies only have a simple bar linking the two door pillars.

prefer. However, as most of the originals lasted many years I am sure new zinc plated steel ones would last quite a while.

BUILDING A NEW BODY.

We are certainly well served by the trade with the T-series cars. A complete new body can be bought for any of the models, either as a kit of parts, assembled but un-panelled, or completely ready to install. If you can afford it, and your body really needs replacing entirely, it is very easy to have a complete body made. If you go this route you usually supply the builder with your scuttle top for him to fit. However, remember that this will need to match up to the bonnet you have. The joint line between the bonnet sides and top should line up with the joint between the scuttle top and the side panel on the new body. If the old body was badly rusted then the edge of the scuttle under the windscreen pillars may need replacing.

If a body is bought without doors make sure it is made so your doors will fit. It may be worth paying for these to be hung - it will save a lot of time. The fit of the rear wheel-arches is important, especially on the later cars where the wings just bolt to them and are not easily adjusted. If the wheel-arches are badly rusted and replacements are needed, take templates of the fit of the old ones in relation to the timber frame and to the top of thc floor of the rear compartment. On a TD/TF mark the position of the rear wing to a fixed point on the body so the new arches can be positioned to line up the rear

A TC body frame as a kit. At this stage the whole job looks very daunting but assembly is quite straightforward so long as dimensions and clearances are checked at each stage as work progresses.

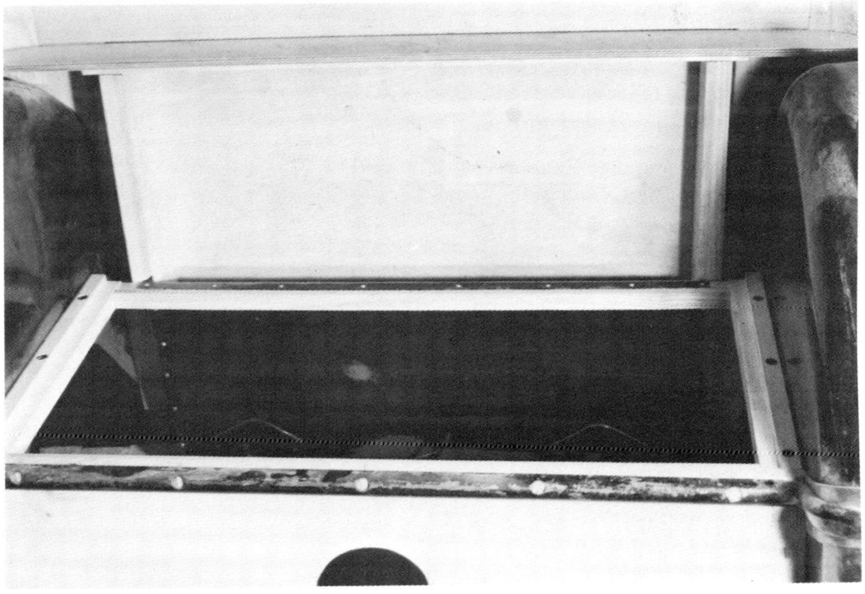

Rear of TC body. The frame for the rear compartment floor is bolted to the body irons and the tie bar. The sides of the sidescreen compartment are screwed to the rear board which is bolted to the body irons. There are two bolts holding the bottom of the sidescreen box to the chassis that pass through the lower rail fixed inside the metal box. The rear board of the body is not vertical, it leans forward a few degrees at the top.

wing with this mark. This is necessary as the captive nuts fixed to the inner wheel arch establish the final position of the outer wing. Incidentally tacking strips for the wheel-arch trim are fixed to the body frame *before* the wheel-arches are fitted - I forgot these on the last TC I built!

When looking at the catalogues to order a kit of parts for the body do not forget that additional metal fittings are needed, if not re-usable from the old body. Also floorboards and sidescreen compartment lid, and the rear body sections for the hood (soft top) as well as the front hood rail. Some of these parts are easily cut from ply at home but if you do want to buy them ready made you might like to order the whole lot together with the rest of the body.

To build a body from a pile of timber, or a kit of parts, will take quite a while. If there is room to keep the old body set up alongside this will help a lot, but in any event take as many photographs and measurements as possible. The best jig for constructing the body is the chassis. Set this up level, packing the stands as necessary. If it is level you will be able to set timbers vertical with a spirit level as necessary, and this will be a help.

Once the chassis is set up, bolt the body frame irons in place assuming these have been cleaned, primed and painted. The TD/TF body irons will also fix the angle of the rear panel but the TA, TB and TC irons are less elaborate. Build up the rear of the body first, forward to the rear door pillars. Remember the hinge face of the

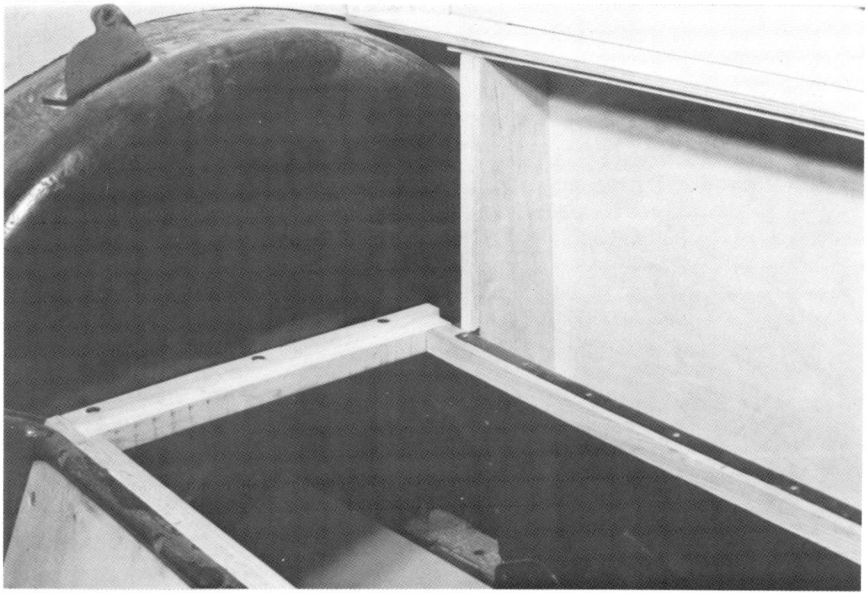

Sidescreen box and wheelarch on a TC. There is a strip of mastic sealer between the rear compartment frame and the body irons.

Checking that the wood is flat before skinning.

rear door pillar is vertical, viewed from the side. The outward lean of these pillars can be roughly established from the rear bulkhead metal sections on the TD/TF and by the body strengthener on the earlier cars. The overall width at the top of the body should be agreed with the measurements taken from the old body.

With the rear section, and rear compartment floor complete, hang both the doors. With the doors in place trial assemble the front sections, lightly screwing or just clamping them in place. It will now be able to be seen if the frame will go together easily and the doors fit. Check the height of the scuttle against the measurements taken from the old body. If you do not have these, consider mounting the radiator in place, and fitting the bonnet tops and sides, to make sure the scuttle is not working out too low. When you are happy you can finally screw the whole lot together. It may help to position the bulkhead on its support brackets to check the height of the scuttle and the fit of the bulkhead to the front rail.

SKINNING THE BODY.

If a new body has been built then you will probably wish to completely re-skin it. The scuttle top should be reusable, which is just as well as this is the most difficult part to make. If you decide to do all the work yourself you should also try to re-use the part of the skinning that covers the curved area behind the doors, even if this has to be cut and welded to make it fit. If working in steel it is difficult for someone without panel beating skills to make a good job of this section unless resorting to cutting and welding to "loose" spare metal. The alternative is aluminium, which is easier to work although half-hard sheet of a heavier gauge than steel should be used. However, if someone can be found who is able to do this part of the job at reasonable cost it may be the best thing to do. The side panel, however, is not difficult to fabricate in either steel or aluminium.

Before starting skinning check the timber frame to make sure that all the joints are smooth where they are going to be panelled over and also that the timbers line up correctly and, on the earlier cars, that the door pillar strengthening plate is in place. Make sure there is sufficient gap at the doors for the metal to fit without causing the door to jam. There are some joints on the body that are not filled or leaded and these are illustrated later. Remember to paint the inside of all panels before fixing them finally in place. If steel is used then it is better to drill pilot holes in the metal before hammering in the steel pins.

Originally all the steel panels for the car were shaped by hand on formers before they were fitted to the completed bodies. The body frame itself is not really suitable for use as a former for working steel panels without risking splitting the wood. It is fine for turning over the half inch edge before securing with pins but any serious hammering should be done off the car. It is particularly important when forming the front edge of the side panel. The timbers the bulkhead fix to are quite thin and these can easily bend in the centre as the metal is worked, giving a curve to the front of the panel. It is

A side panel for a TC as supplied ready to fit. However, this is not a difficult panel to make in the home workshop.

A strip of steel is clamped to the face of the door opening and the shaped panel to the outside of the body. Thick plywood cut to the shape of the door opening is clamped to hold the outside panel firmly against the wood. The edge is gently turned over onto the inner strip.

Once the edges are turned over the outer panel is welded to the inner through small holes drilled in the edge of the outer panel.

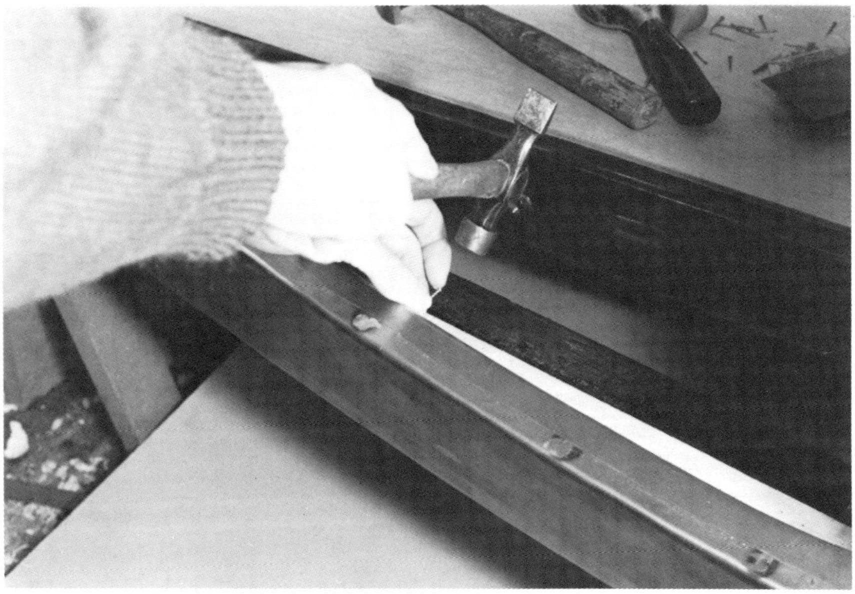

Once the inside of the panel is painted the edges are turned over inside the body and pinned in place. It is easier if small pilot holes are first drilled.

better to remove the work from the car, after marking a line for the timber, and make the fold at the bench. Remember, however, the metal will need to be shrunk a little as it is folded over to curve the panel as viewed from the front. With aluminium sheet it may only be necessary just to wedge the centre of the front rail against the door pillar to stop it bending. If using aluminium, soften the edges occasionally by rubbing the surface with household soap and heating with a gas torch until the soap blackens.

TAs, TBs and early TCs just had the side panel turned over the wood and pinned, whereas the later cars all had three separate pieces of metal fixed round the inside edge of the door frame. The side panel was welded to these and the strips were then turned over the inside edge and pinned to the wood, making a much neater looking job. The easiest way to do this is to clamp the strips of metal around the frame first. Clamp the side panel in place using a piece of plywood shaped to the size of the panel to hold it firmly all along its edge against the body frame. Gently work round the lip turning over onto the strips of metal round the door opening. When you are happy with this drill, or punch, a series of 5 mm holes through the edge you have just turned over and weld through these with a MIG welder to simulate the spot welds of the original panels. After removing to paint the inside, the side panel is once again clamped in place and the inside edge turned over and pinned. Remember to apply some primer to the edges of the side panel where it is covered by other panels.

I have talked here about using a MIG welder. I find that having one of these around the workshop makes things a lot easier. Even if large welding jobs are passed on to professionals there are so many uses for a welder that the cost of buying one is soon recovered. The small D.I.Y. welders now available are adequate for most needs and are easy to use.

THE DOORS.

So far I have not mentioned the doors. Possibly the memory of my last rebuild still lingers! The T-type door is not the easiest item to work on. The wood has compound curves in all the pieces, and it is very difficult to fit them into an existing door skin if just the wood is being replaced.

To replace the wood in sound door skins. First carefully remove the fixing pins and fold back the turned over edges. Remove the old wood for copying or comparing with purchased replacements. Assemble the wooden

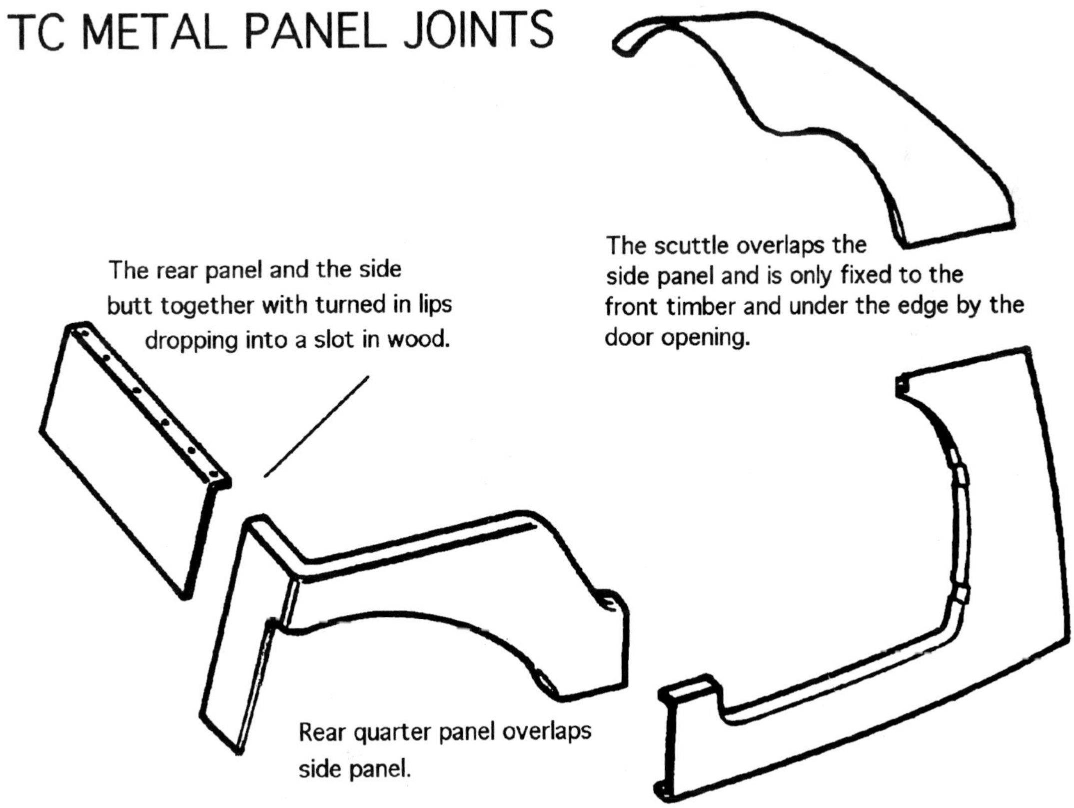

A thick plywood former is clamped over the metal skinning before the edges are turned over.

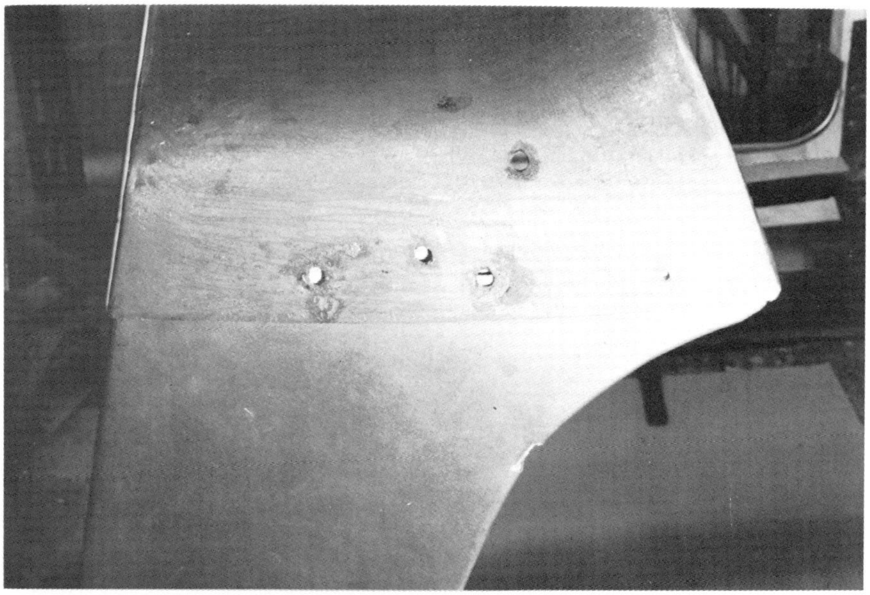

The scuttle overlaps the side panel and is fixed to the frame only at the front and under the edge by the door opening.

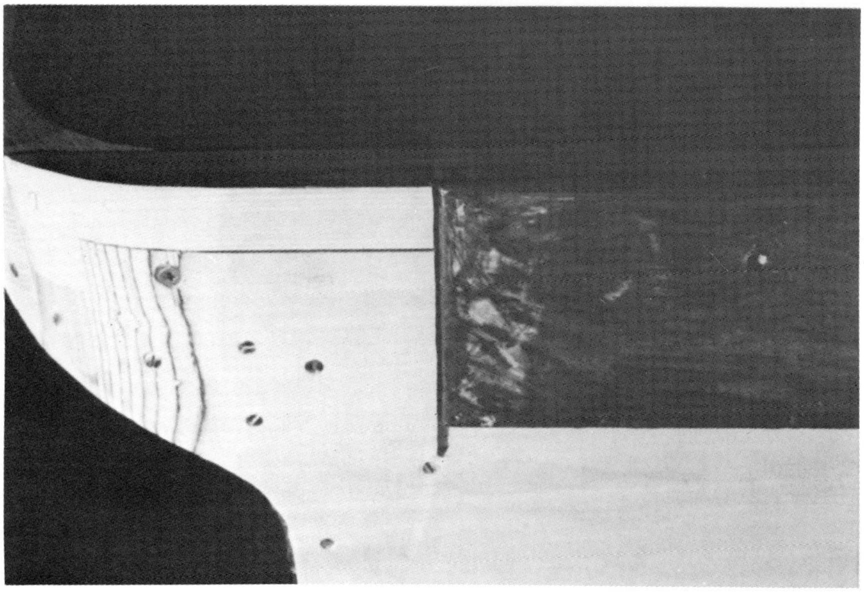

The edges of the rear quarter panels and the rear panel fold over and slot into a recess in the wooden back panel of the body frame.

frame and carefully check against the old wood and the door opening in the body to make sure it will fit and that the twist in the frame is correct. If not get this right before fitting the pieces into the door. Dismantle the frame and re-assemble them within the door, making sure that the tapped plate for the sidescreen bracket is in place. The front piece goes in first, followed by the bottom and the piece by the hinges. With these three in place lock the hinge end of the top piece into the frame and force the front of it to the top. If the wood is the right size it will all go together well. If it does not then it is back to measuring and trimming to fit. When the wood is in place re-fit the strengthening piece at the hinge end. The bracing strut should be fitted once the door is in place and correct. As the bolts holding the hinges to the doors and door pillar can slacken off after a while it is better to use self locking nuts to hold them. Whilst talking about fastenings, I should mention that before driving screws into new wood it helps to dip the end of the screw into some grease to lubricate it.

FLOOR BOARDS.

As mentioned earlier the floor board can either be purchased from suppliers or cut from 9 mm exterior grade birch plywood, using a jig or band saw. Either way check the fit of the boards before painting and installing them. Too often the floor boards are fitted before the trim is installed only for the owner to find at a later date that the trim has to come out before they can be lifted to attend to some mechanical problem. The floor boards must be small enough to slide past the side trim panels when required. They are fixed in place by a combination of machine screws and wood screws that varies from model to model. Tee nuts are fitted to the floors of the TD/TF for both the gearbox cover fixing machine screws and for the seat runners. Later L.H.D. TDs and TFs had a metal tray let into the floorboard to give more space for the driver's feet. A rubber mat sat in this well. On the TA, TB and TC there is a wooden floor board packing strip fixed along the edge of the

TD METAL PANEL JOINTS

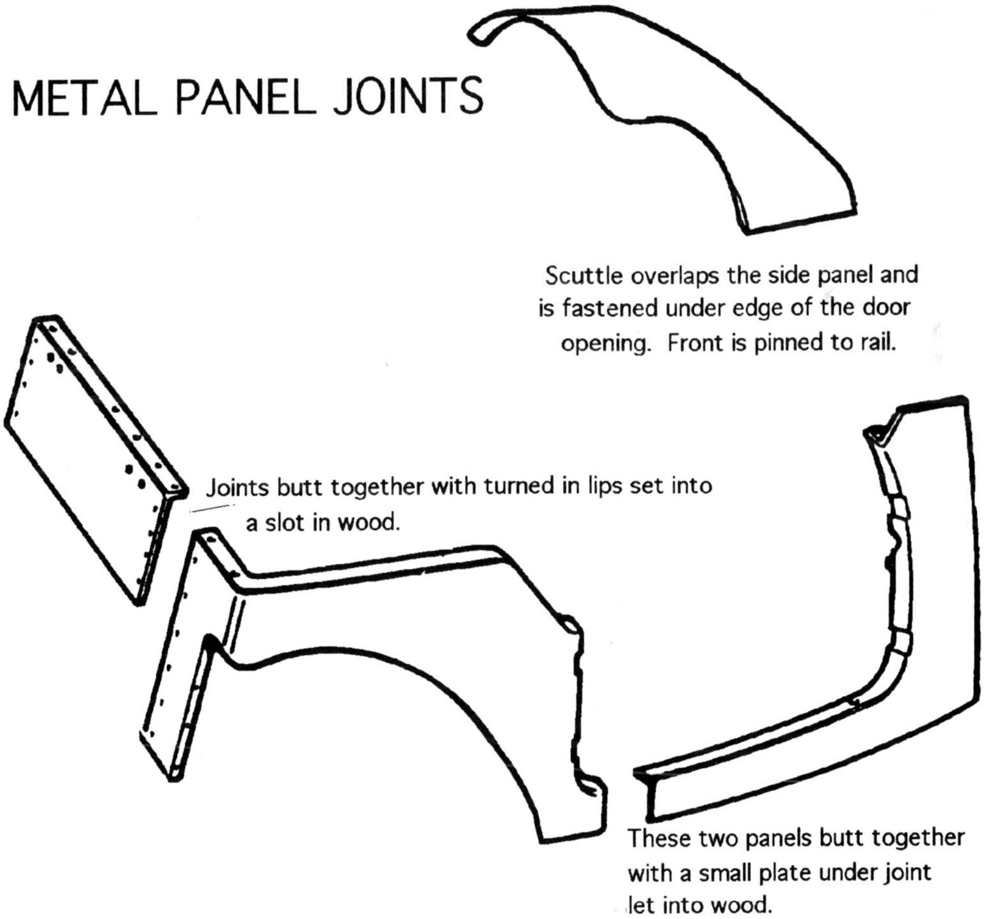

board where it meets the prop shaft tunnel. The floor boards will probably not be finally fixed until all the wiring etc. is complete and access to the chassis is no longer required.

Before final fixing, the boards should be painted with a flat black paint - I use a "blackboard" paint for the top surfaces and an underseal for the lower surface. Pay attention to sealing the edges against draughts and water. Mastic strip is a useful sealer and this is available in packs sold for sealing rainwater gutters on houses. On the TD/TF there are pieces of rubber about 3 mm thick that fit around the ends of the scuttle bracing tube, under the floor boards. There is also a rubber seal that fits on top of the pedal box, also under the floor boards, in addition to the bellows seal around the pedals.

FINAL ASSEMBLY.

Once all the body is finished and the wings and other panels repaired the whole car must be assembled to check fits and to drill any necessary holes before the parts are painted. It may seem to be making a lot of extra work but it is a lot easier to fit the car up without worrying about damaging the final paint finish and if any alterations are required it is better to do them at this stage.

So that the wing piping fits snugly when the car is finally assembled allowance must be made at the trial assembly stage for the thickness of the piping. Either the piping itself will have to be fitted or scrap pieces of P.V.C. used in the joints in its place.

Mastic sealer sold as a continuous strip for use on house guttering is very useful for sealing floorboards.

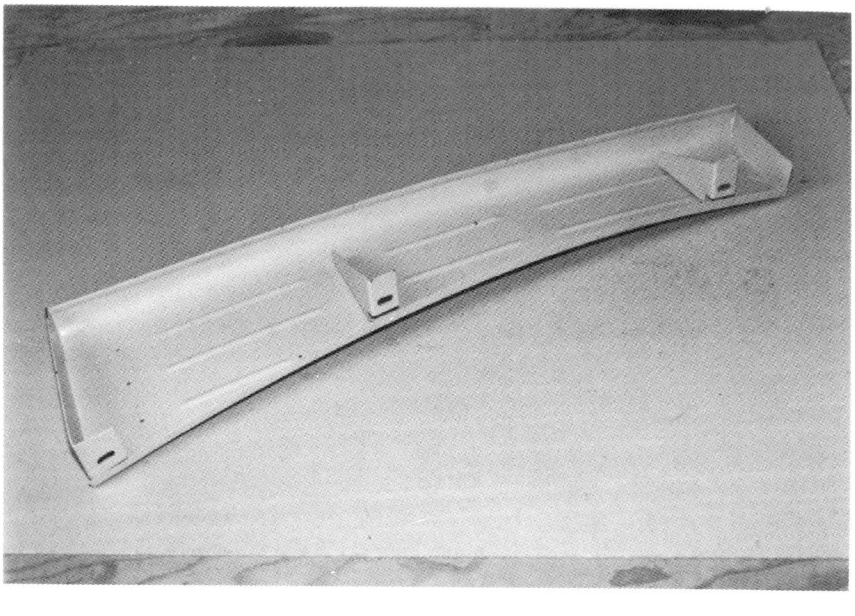

TC running board. This is an original factory panel. Some of the reproductions available do not have sufficient curvature.

This rather second-hand looking TD bulkhead was restored to a good state by a skilled panel beater.

The same panel after restoration.

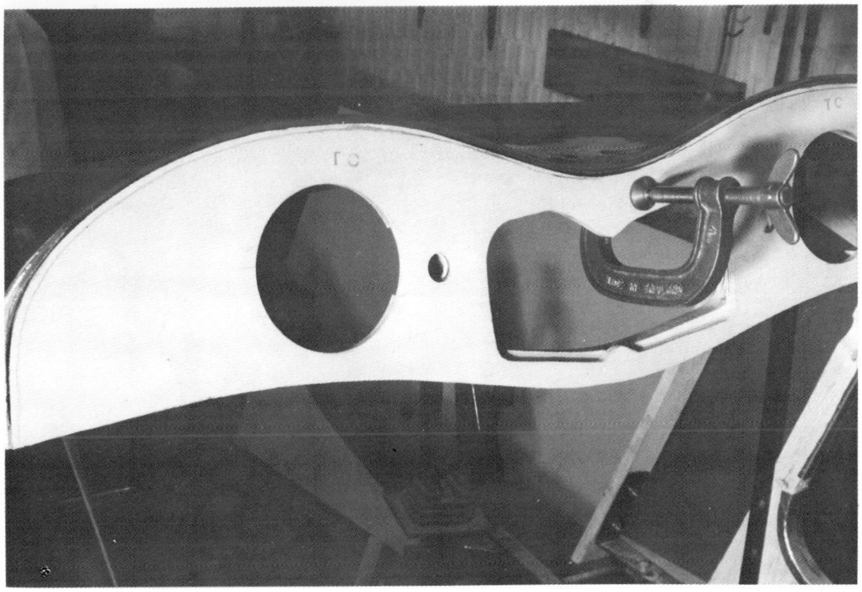

The dashboard must be tried out against the body before covering and cut, if necessary, to allow room for the piping and chrome trim.

A TD body and scuttle being fitted as a unit.

CHAPTER FIVE

THE ENGINE

The engines fitted to the TA were very different from those in all the later T-series cars. As the later engines were more numerous the bulk of this chapter deals with them but there is a section where the special problems the earlier engine poses are considered. However, the general principles of engine re-building will apply equally to either type so the following sections will be of some use to the TA owner.

THE XPAG AND XPEG ENGINE.

The XPAG engine fitted to the TB, TC, TD and TF was a simple, but strong, engine. It was extensively used by specials builders as well as by constructors such as Lotus, in the Lotus VI, Cooper, Lester and Keift. For a period in the early 1950's it seemed that everyone was working to produce even more power from this unit. There were a number of specials builders who developed this engine to provide the power for very sophisticated chassis, most of them much lighter than the original M.G. item. It follows, therefore, that, if carefully rebuilt, an XPAG engine in a T-Type used on the road is capable of long and trouble free service. There are a few weaknesses we will look at later but none that should make the car difficult to run or maintain.

DETAILS OF CHANGES IN PRODUCTION.

The first XPAG engines were fitted to the TB and differed from the TC engines only in the triple valve springs and lack of a timing chain tensioner. The 1,140cc Morris 10 engine the XPAG was developed from had a different head and weaker rods. If nothing else is available the block from this engine could be bored out to 66.5mm to give a capacity of 1,250cc, provided it was good block. There were no significant modifications to the TC engine during the production run of the car.
Engine numbers:- XPAG 501 to XPAG 882 are TB.
XPAG 883 onwards are TC.

The TD model introduced the following changes to the engine:-
The starter motor changed and a new ring gear was fitted. There were 93 teeth on the old ring and 10 on the starter pinion, 120 and 9 after. To fit the later starter on a TC you would need a new ring gear. The drain tap for the block was moved from its central position to the front of the block. The air cleaner and pipe work was changed and modifications made to the carburettor linkage to combine the mixture enrichment and slow running controls. The fan gained a spacer and the pipe work for the thermostat housing modified. The rocker cover was slightly changed.

There were a number of modifications to the engine during the production run of the TD and these are listed in a previous chapter. The major change was the larger

TF ENGINE

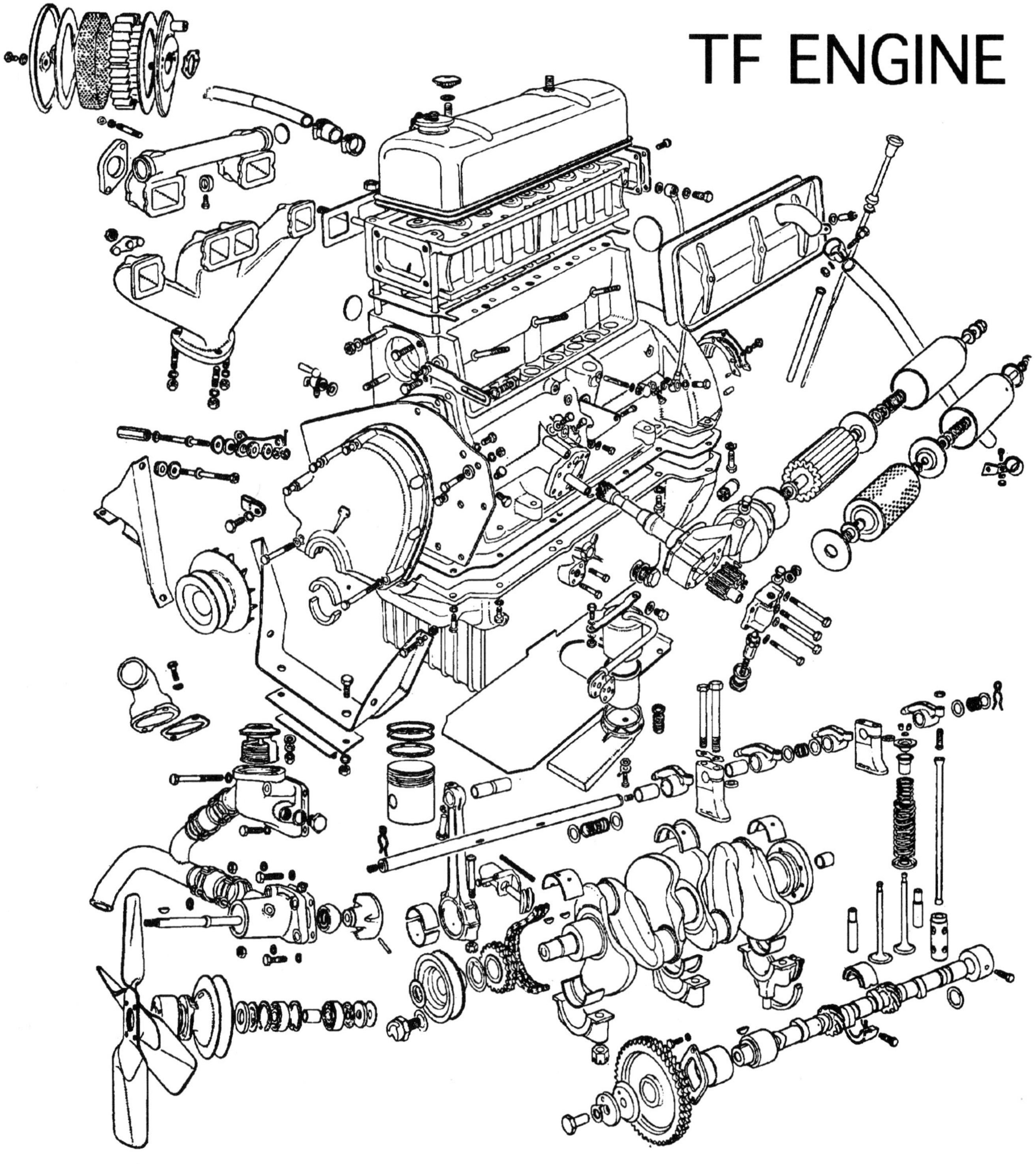

clutch, fitted from engine number 9408, at which point the prefix was changed to TD2. The bell housing and clutch release fork were also changed, as was the flywheel. At engine number 14224 the oil pump was drastically changed for the better. The external oil pipes were dispensed with, removing a major source of worry to the "press on" T-type driver - a fractured oil pipe could empty the sump in less than 60 seconds! Engine numbers:- XPAG/TD/0501 to XPAG/TD/9407, XPAG/TD2/9408 to XPAG/TD2/30300. From XPAG/TD3/17029 the Mark ll engines carried the TD3 prefix within the TD2 sequential numbering.

With the introduction of the TF, the engine gained the larger valves of the TD3 Mark ll engine with a lower compression ratio of 8.1:1. The carburettors were the larger 1 1/2in. size, but with a different body casting number. These were fitted with pancake air filters. The engine was later modified to enlarge the bore to 72mm increasing the capacity to 1,466cc. Previously some brave souls had managed to overbore standard 1,250cc blocks to this size, but usually had to fit liners. The factory modified the block casting to overcome this problem. The connecting rods are heavier and a new head gasket used. The compression ratio was increased to 8.3:1.
Engine numbers:- XPAG/TF/30301 to XPAG/TF/36330, XPEG/TF/0501 to XPEG/TF/3940.

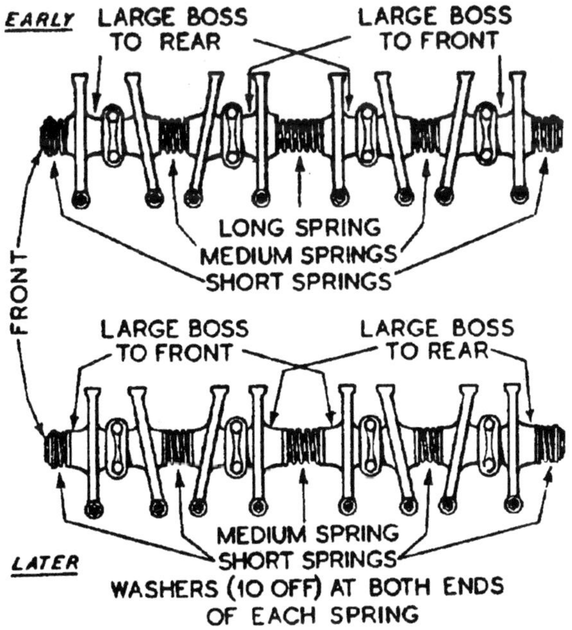

EARLY AND LATE XPAG ROCKER SHAFTS

NON T-SERIES ENGINES.

XPJM and XPJW. These 1,140cc units were fitted to the Morris 10 and Wolseley 10/4 respectively and, as mentioned earlier, have different cylinder heads and rods from the XPAG. Externally they lack the cast-in M.G. octagon. You may find a block bored out to 1,250cc as a replacement for a damaged XPAG one and if this is the case do not try to re-bore to an oversize.

XPAG/SC. This is the single carburettor version of the engine and was fitted to the Y and YB saloons.
Engine numbers:- XPAG/SC/10001 to XPAG/SC/16915.
At the same time the TD2 engine was introduced the YB gained a similar SC2 version.
Engine numbers:- XPAG/SC2/16916 to XPAG/SC2/18460.

XPAW This 1,250cc unit was fitted to the Wolseley 4/44 from its introduction in 1952. Except for the location of the dipstick the block is almost identical to a later TD. This makes it a very useful item for the T-type owner. In addition to moving the dip stick, the sump will need to be changed and the oil pipe fitted to the rear main bearing cap will have to be lengthened.

WEAKNESSES OF THE ENGINE.

As I said, this a basic, simple and strong unit and easy to rebuild and maintain but there are a few weak points.

The engine has a relatively long stroke by modern standards and this, allied to its free revving nature and fairly low axle ratios, means that the piston speeds are high. This high piston speed causes wear and a re-bore may be needed rather sooner than on some engines, and piston rings can break up. This is not, however, going to trouble most owners today when average annual mileages are low.

The valve gear is very prone to wear but this seems to vary vastly according to the use the car receives. Cars that are used only rarely can wear out camshaft lobes and cam followers at quite low mileages and perhaps this could be due to the oil leaving these areas during lay ups. However, any engine I strip seems to have a lot of oil everywhere so I am not sure this is the cause. Another theory centres around the time it can take for oil pressure to build up after a long period out of use and the only thought on this I can offer, is to drain off a couple of pints of oil and add some fresh oil just before starting up. This only after a break of months not days! The idea is for the oil to run down the valve gear to provide fresh lubrication in this area. Most of the wear on a new camshaft occurs when it is first started and to help avoid this you must use an engine assembly compound like "lubriplate", or one of the compounds supplied by camshaft suppliers, when re-assembling an engine. Make sure the pump is primed before starting up and, as soon as oil pressure is seen, take the revs. to 1,500 plus, do not run the engine at idle at first, the higher engine speed will reduce inertia loads. There is a case for reducing the load on the valve gear by using modern, higher rate, valve springs. Some specialists will be able to advise and Brown and Gammons can sell you a kit to enable some of these springs to be fitted. Autojumbles can sometimes yield Morris 10 springs that will fit - valve bounce will be 500 revs. or so lower, but this will not hurt for road use.

If all this sounds alarmist, I am sorry! You should not let any worries over the valve gear spoil your enjoyment of the car - if worn it will only cause a few more rattles in that area and a slight fall in performance. Obviously when *very* worn overheating could result, because of less efficiency, but this will take some time to happen.

The last weakness is a tendency to leak oil. This will drip from the front pulley, around the joint with the timing chain cover and sump, the rear main bearing through the bell housing and from all those external oil pipes. Care on assembly will help but unless some major modifications are done you will not eliminate these entirely. Buy a couple of drip trays.

The oil seal at the rear of the engine consists of threads machined on the crankshaft combined with an oil thrower. The effectiveness of this arrangement relies on a fairly close fit between the crankshaft and the rim of the main bearing cap and the crankshaft and an aluminium cap fixed to the block. Both the bearing cap rim and the aluminium slinger cap can be worn if the engine has run for any length of time on worn main bearings. The aluminium cap can be replaced but the bearing cap would need to be built up and re-machined to size. This would need to be done with the cap mounted in the block to ensure it was concentric with the crankshaft.

The aluminium cap is dowelled to the block and fixed with three screws. If it is re-positioned or a new cap fitted it will need re-dowelling. It is difficult to measure clearance between the crankshaft and the slinger or bearing housing. As a rough guide I have used a piece of paper, about 2 thou. thick, wrapped round the crankshaft to space the aluminium cap away from it before marking its position. Care must be taken to seal the aluminium cap to the block and also to put some gasket sealer on the threads of the bolts holding it - one of these is threaded close to an oil way and leaks along the threads can result. The oil drain in the bearing cap should be clear and care must be taken when fitting the cap to seal the joint with the block with gasket sealer without applying too much - it must not form any obstruction that could stop the oil getting away down the drain tube.

The rope seal at the front of the crankshaft, where the pulley is attached, often leaks. It is easy to replace this with a modern lip seal, without any modification to components or machining. I have used a seal 47mm x 36mm x 7mm, rubber cased, set in the housing with "instant gasket". The housing must be cleaned thoroughly to remove any oil.

Finally overheating! There are many stories around about these cars overheating but usually this is a symptom of other problems, both of modern fuels and of mistakes made when rebuilding or setting up the car. The TB, TC and TD are probably over-cooled and the only trouble usually encountered on a good car is one of fuel starvation, owing to heat from the exhaust manifold affecting the carburettors. The TF has less space around the engine and can run rather hot. The water circulation around a TA engine is not perfect and these can also run hot. We will look at all these problems later.

BEFORE YOU START THE REBUILD.

When the car was bought you may have been lucky enough to be able to drive it before starting on any rebuild. If so, you will have had a chance to assess the condition of the engine, and to decide if it is in fairly good health or not. If wished a compression check can be done. A gauge is not expensive and turning the engine over 5 or 6 times on the starter for each cylinder, with the plugs out and the throttles open, will give some readings. If there are one or two much lower than the rest, try again with some oil around the piston rings, introduced through the plug hole, and if the reading is still low the trouble will be a poor valve seal. You will

Later XPAG engine. Note the oil pump with integral filter.

probably still wish to strip the engine but at least you will have some idea if a re-bore is needed, or if just a valve job will do. I hope that once the engine is stripped you at least look at all the areas we will talk about even if the engine is just cleaned and reassembled.

About cleaning and dismantling. Do not touch anything until you intend doing the whole job. If the time or funds are not available to rebuild the engine right away then leave it in one piece. Just ensure every drop of water is out of it - even taking out the core plugs in the block if you are not sure. If stored in a cold area with even a small amount of water in it, there could be some expensive repairs necessary later. Pour a quantity of oil into each cylinder and turn the engine over before replacing the plugs. This will help prevent rust and soften the carbon on the piston crowns.

When starting to take the engine apart clean off the outside first, steam clean if you like, but once cleaned dismantle the rest at the same time in case any water has found its way into the internals where it could cause rusting. Keep ALL the bolts carefully labelled in bags. All the threads on the XPAG engine are metric, 8mm x 1mm, which is not of the same pitch or profile as those now used. It may be possible to replace these later from one of the specialist suppliers, but some are not available. Remember, however, that bolts may have been used in the wrong place by a previous re-builder!

As the engine is stripped always store all the components in such a way that they will be able to be replaced in exactly the same position. For example each main bearing cap must fit the same place and the same way round as before. The bearing caps bear "diamond" symbols and numbers and a note of these should be made as an aid to identification. If you are only going to clean and re-lap the valve gear you must have each valve going back in its original place with the same springs and push rod. The connecting rods are marked and, as with the rest of the engine, make notes of which bore they fitted. Over the years the engine may have been rebuilt many times and components may have been replaced or fitted to the wrong place or the wrong way round. A common fault is for the rods to be fitted to the pistons with the oil feed holes pointing the wrong way. Rapid bore wear results! Look carefully for the timing marks and for their position in relation to the timing chain.

STARTING TO STRIP THE ENGINE.

Assuming the engine has been removed from the car, and it is safely standing upright in an area where the odd drop of oil coming out will do no harm, start by removing any ancillary brackets and fittings not already removed and store these in labelled bags - leave the bolts with them for identification. Remove the clutch, marking the cover and flywheel if you know that these have been balanced as a unit. The flywheel cannot be removed from the crankshaft until the sump is off. Remove the rocker cover and valve gear and pack away carefully. Then the cylinder head should come off, but not dismantled further, followed by the water pump, distributor, oil pump and all external pipe work. Take care with the oil pump bolts - they snap very easily, leaving a piece to be extracted from the block. Remove the push rods, the side cover plate and the cam followers.

If a complete rebuild is being done then the cylinder head studs will have to come out sooner or later, and this may be the best time to remove them. The easiest way, if you do not have access to a stud remover, is to lock two nuts together and just wind the stud out with a ring spanner. This usually works very well and will not damage the stud. The engine can then be turned upside down, but the face of the block must be protected from damage. If using a clean, smooth surface just have a piece of card or similar to act as protection should the block move whilst working on it.

The front pulley will have to come off next. This will either be easy, or it may take a while! If you find that you cannot undo the starting handle bolt then you may have to remove the sump first so as to be able to wedge one of the webs with wood, such as a hammer handle, so that enough pressure can be applied to the bolt. Take care not to crack the timing chain case if forcing the pulley off with a couple of tyre levers. Try to apply the force to the head of the securing bolts rather than just anywhere on the casing. Examine the pulley for signs of cracking of the keyway and for wear on the seal face. Remove the sump and the timing cover and look for the timing marks to see if they appear to match the correct pattern shown in the manual. The engine will have to be turned over until the two gears are in the correct orientation. Mark the links with white paint, as shown, in case the chain is re-used although the links should always be counted on re-assembly to ensure the chain is in the correct position. Remove the tensioner, storing carefully, and, after removing the bolt on the end of the camshaft, ease the timing gears off the crankshaft and camshaft.

There are now only the crankshaft, pistons and camshaft to remove. The camshaft needs care as it will have to be freed off the centre bearing and pulled out along with the shaft. The bearing splits in half to remove. Try not to damage the front bearing as this will have to be line bored if it is replaced. It may be possible not to need to disturb the plug covering the end of the camshaft if only the centre and rear bearings are replaced. In any event it is worth mentioning here that the fit of this plug is vital - if it falls out at any time all the oil will be quickly lost into the bell housing. If it is necessary to remove the plug give the camshaft a couple of taps with a soft faced hammer, after removing the bearing securing bolts. This should remove the plug which is 45mm in diameter, a different size to those used elsewhere in the block. To remove the crankshaft and pistons, I usually release the big end caps first, keeping the bolts, nuts and caps safely lined up where I can identify them, then undo the main bearing caps and lift out the crankshaft and flywheel together. With the crankshaft out of the way, return the caps to the big end bearings to keep these together, and carefully extract the pistons.

EXAMINATION OF THE ENGINE.

This is perhaps the one area where you may need to call on some assistance if you do not have the measuring tools and knowledge to use them. Before starting cleaning take a look at the bores, examining them for signs of scores and for a ridge at the top.

If there is any doubt about the condition of the bores now is the time to either measure them yourself or take the block for an expert opinion. You will have to decide here if you are going to approach machining companies direct, or through an M.G. specialist used to working with XPAG engines. My advice would be that unless you know that the company you intend using has a good reputation, and works with older engines, it may be better to entrust the sub-contracting of any machining to an M.G. specialist. True you may have to pay for his time, but he will know the good people to use, and you will have someone to shout at if anything goes wrong! Mind you there are good and bad M.G. specialists, so ask other owners first.

Anyway we are rather straying from the point here. You are faced with all the parts of your engine and are trying to decide just what needs work. Apart from the condition of the bores themselves, how about the pistons? If the bores are bad then forget about the pistons as new ones will be needed anyway. However, if the bores are good, look at the pistons to make sure the rings are unbroken and there is no sign that the crowns are damaged by overheating. If the engine was left for a while with oil in the bores, the carbon on the piston crown should scrape off easily but care must be taken not to score the metal. It is easier to clean the top of the piston before removing it from the bore, but take care where the removed carbon goes if you are not doing a complete strip down. Look at the piston rings. Are any broken? If there are some broken rings then it is likely the engine has done quite a mileage since the last re-build. It is very difficult to obtain rings separate from pistons now although a few sets do turn up occasionally. Remember if you do re-ring an engine, or fit new pistons without a re-bore, the surface of the bore will need to be "glaze busted" or honed to allow the new rings to bed in. Sets of rings sold for fitting to partly worn bores usually have "stepped" top rings so that they do not hit any slight ridges left at the top of the bores by the old rings.

We will look at the pistons again shortly, but what of the other parts that will need work. Firstly look at the crank and the bearings. You will get some idea of their condition by looking at them but if in any doubt have the crankshaft measured and look at the surface of the journals for signs of scoring or bad machining. Unless you know that the crankshaft is new you *must* have it crack tested. I took what looked like a perfect crank for testing and was horrified to find it was cracked almost all the way through! Whilst we are talking of crack testing, you should also consider having the rods tested.

If the engine was bought as a "box of bits" and it has not been seen to hold water, consider having the block and head crack tested before spending a lot of money on re-building them.

As mentioned earlier, the valve gear is often worn and, if the engine has seen much use, it is probable quite a few parts will need to be renewed. The camshaft and cam followers are usually replaced as a set, although I have found that sometimes only one cam follower is worn and this may just be because of a fault in the hardening of that follower. Measure the lobes of the camshaft to see if the wear on that follower is matched with wear on the mating cam lobe, as might be the case.

Luckily camshafts are available, some with improved valve overlaps to improve performance. I would not, however, go for a "hot" camshaft for road use. Take advice on tuning from specialists, such as Brown and Gammons. I will include some details of modifications later. Unless it looks very good replace the timing chain. The timing gears are also available. The rocker arms and rocker shaft should be examined for wear. The rocker arms wear at the tip and can be re-faced by grinding as long as this is not taken beyond the surface hardening.

In the camshaft area is the oil pump. Strip and examine for wear. I will talk more of this later but take a look at the cover early on in case this has been badly scored. Wear is usually a sign that the gears are past their best. It may also be a sign that the gears have been turned over at some time. If worn gears are turned over then the wear pattern on them will tend to force them outwards, acting as an end mill against the cover, and sending a stream of metal swarf round the engine. The golden rule is not to reverse the gears in an attempt to equalise the wear, replace them if worn and at the same time replace the removable dowel holding the non-driven gear. Looking at the pump early on will give some idea what the rest of the engine may be like and also, if it is necessary to have the end cover ground to remove wear, you could send it with the other parts to the machinist.

HAVING MACHINING WORK DONE.

Right, so you have looked at the engine and decided that new main and big end bearings, re-ground or new crankshaft and a re-bore and new pistons are needed. Before rushing off to ask the machine shop to re-bore the block, look to see what there is available in the way of pistons, bearings etc. You will find that not all sizes are available at any one time and you may have to compromise. If the engine has already been bored out to the maximum oversize then liners may have to be fitted. If liners are fitted it is nice to have the bore cut in the block with a step at the bottom for the liner to sit on, so it cannot be forced down in use. The only disadvantage of this is that it is a lot more difficult to fit the pistons and rings from below unless the inner bore of this step exactly matches the size of the inside of the liner. On one of my engines this is of a slightly larger bore than the liner, which will usually be the case, and the piston rings come out of the clamp and jam against the lip of the liner. The only remedy is to make a special thin piston ring clamp from some shim stock and a large jubilee clip. Not a nice job!

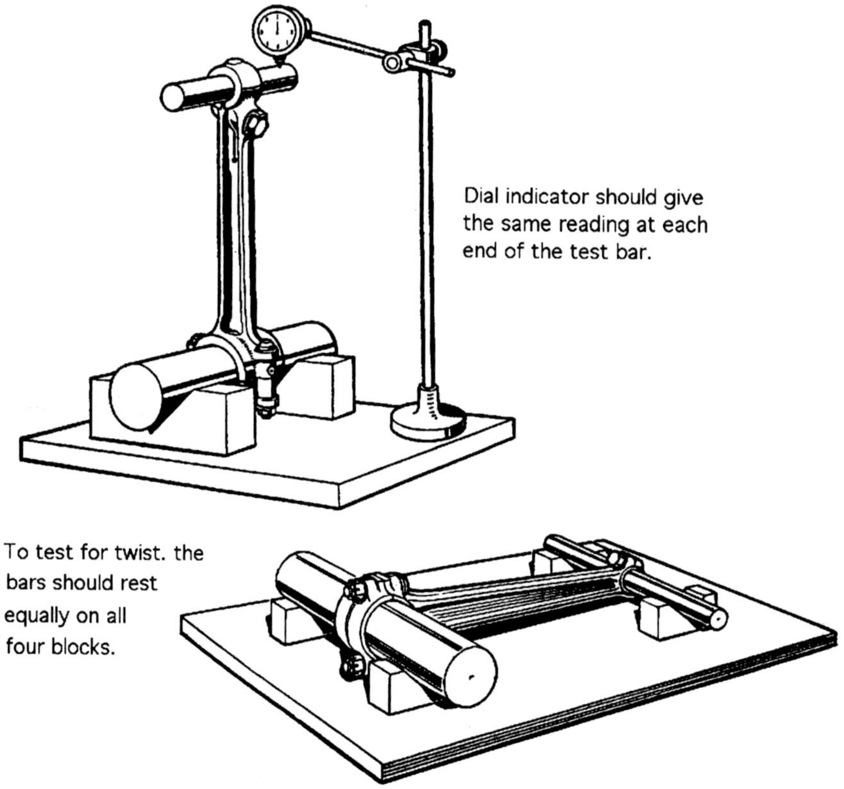

CHECKING FOR BENT OR TWISTED RODS

Do buy the pistons before the block is bored. The machinist can then have them to check the dimensions, in case they vary slightly from standard specifications.

The machinist will also be able to arrange for all the parts to be balanced. The rods on XPAG engines varied during the production run, getting progressively heavier. There was a code cast into the rods and it is easily seen if there is a matched set. If the weight varies slightly it is possible to have them balanced by removal of small amounts of metal. If they vary greatly some weight will have to be added to the lighter ones, or else look for some other rods. The rods, pistons, pins and bolts should be balanced as sets - although I have seldom found a lot of difference in the weight of a set of gudgeon pins or small end bolts! Have the clutch cover, flywheel, and crankshaft balanced as a unit. Mark their relative positions with white paint. Whilst the flywheel is at the machinist you might like to have some metal removed to reduce its weight - it is up to 30% heavier than it needs be.

Whilst talking about rods, it is as well to be aware that these have often been damaged by previous owners. One favourite trick of some "old school" mechanics was to try to remove the symptoms of worn out big end bearings by filing the bearing caps. This means that the nice new bearings you try to fit will be too big. The only remedy is for the rod and cap to be re-bored to the correct dimensions - this will remove some metal and will affect balance, as well as possibly piston height. The rods may also have been bent or twisted - a method of checking for this is given in the TD/TF manual and reproduced here for TC owners. A less accurate method to check for gross errors is to fit the crankshaft and assemble the rods on the crankshaft, without the pistons fitted. Rotate the crank and ensure that the rods are central in the bores.

CLEANING ALL PARTS.

I am probably getting a little out of step here as most of the work mentioned above will be done once the parts are clean. However, I will just say here that it is necessary to carefully clean all parts before re-assembly, particularly after any machining is done.

Starting with the block. Remove all core plugs by punching a hole in them with an old chisel, or similar, given a good edge on the grinding wheel. The plug can then be levered out. DO NOT bang away too hard at the plugs and crack the casting of the block - especially easy where the cylinder wall is just behind a plug. The whole block can be chemically cleaned in a caustic solution or just cleaned using a wire brush, or similar, to remove paint outside and various brushes, rods and cloths to clean the rest. All oil-ways must be cleaned and particular attention paid to the water passages to ensure they are clear. On the side of the block there is a cast water passage with 35mm core plugs at either end. With the core plugs removed make sure that the passage is clear and that there are no lumps of casting blocking the flow. Then, AND THIS IS VITAL, look for either one or two 5mm diameter holes passing into the main water jacket. The early engines seem to have only a hole at one end, the later engines at both. The hole is often blocked with sediment but it must be located and cleared, using a rod or drill bit.

Great care should be taken to clear all oil-ways in the crankshaft, especially after machining. Also look at the oil holes in the rods, rocker arms and shaft and all external pipes. A clean workshop is essential as it is all

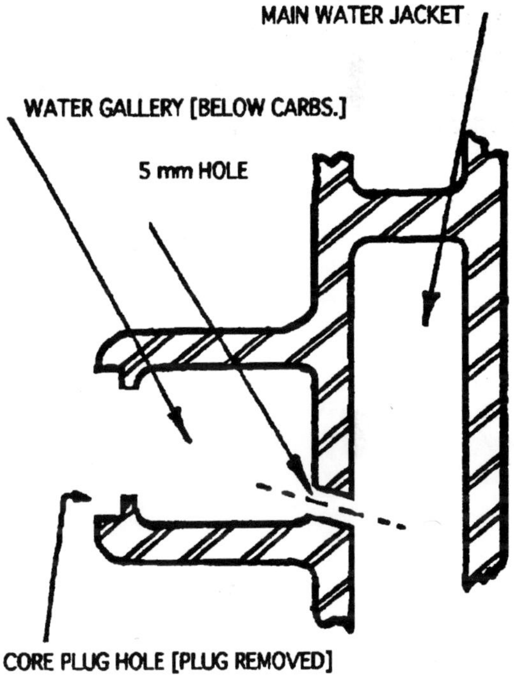

too easy to transfer dirt to oil-ways when working on other parts.

THE CYLINDER HEAD.

I have paid no attention to the cylinder head so far. When you remove the valves examine them and also look at the seats. It may be that only a clean and grinding-in will be necessary. If so you will be lucky, but the most usual thing on high mileage engines is for the valves to be pitted and the seats eroded. Modern unleaded fuels can cause valve seat erosion, although I have been told that many owners in the United States run their cars for high mileages on unleaded fuel without much harm. It seems that erosion occurs at a greater rate at higher engine speeds and it is possible to fit hardened valve seat inserts, bronze valve guides and valves with chromium plated stems to prevent this erosion. However, fitting inserts in an otherwise sound head perhaps adds to the chance of engine failure should a seat come loose so maybe it would be better to leave well alone until the need to replace the valve seats arises.

THE OIL PUMP.

Full details about clearances etc. are given in the manual but basically you are looking for an end clearance, the gap between the cover and the gears, of around 2 thou., and a clearance between the gears and the housing of 3 - 4 thousands of an inch. The bushes in the housing should be good, as should the driven gear shaft. This shaft is easily replaced. There is a tapped hole in the end and a bolt can be screwed in to allow a ring spanner to be used to turn the shaft, which is a press fit in the block.

There is a paper gasket between the pump body and the block, but none between the cover and the body. The drive gear and shaft are keyed to the pump gear and secured with a circlip - the ears of this clip must be removed. The oil pressure relief valve ball is often worn and may need replacing. There is a bypass valve mounted in the block under early pumps but this was moved to the pump itself on the later type with the built-in oil filter. The by-pass valve is there merely to allow oil to flow, unfiltered, round the engine if the oil filter is left so long between changes it blocks up! We will look later at any changes needed to fit the later type of pump to an early engine.

The pump must be assembled to the block before the crankshaft and timing chain are fitted, so you can feel

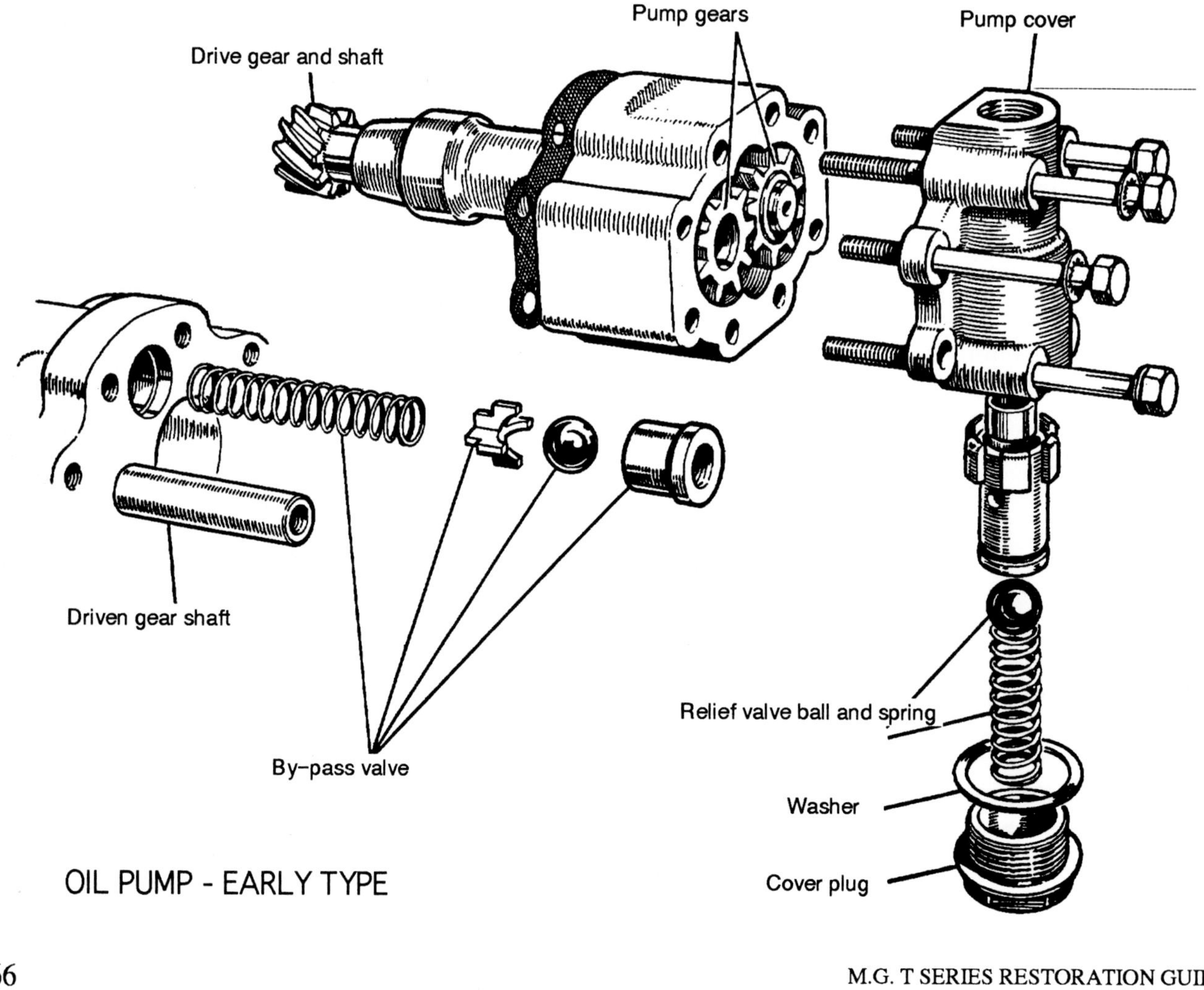

OIL PUMP - EARLY TYPE

Comparison of oil pumps. The TA pump is shown above. On the left is the type fitted to the TB, TC and early TD. The pump pictured below, with the integral filter and no external pipework, was fitted to later TDs and all TFs. The external pipes can fracture leading to rapid loss of all the oil in the sump.

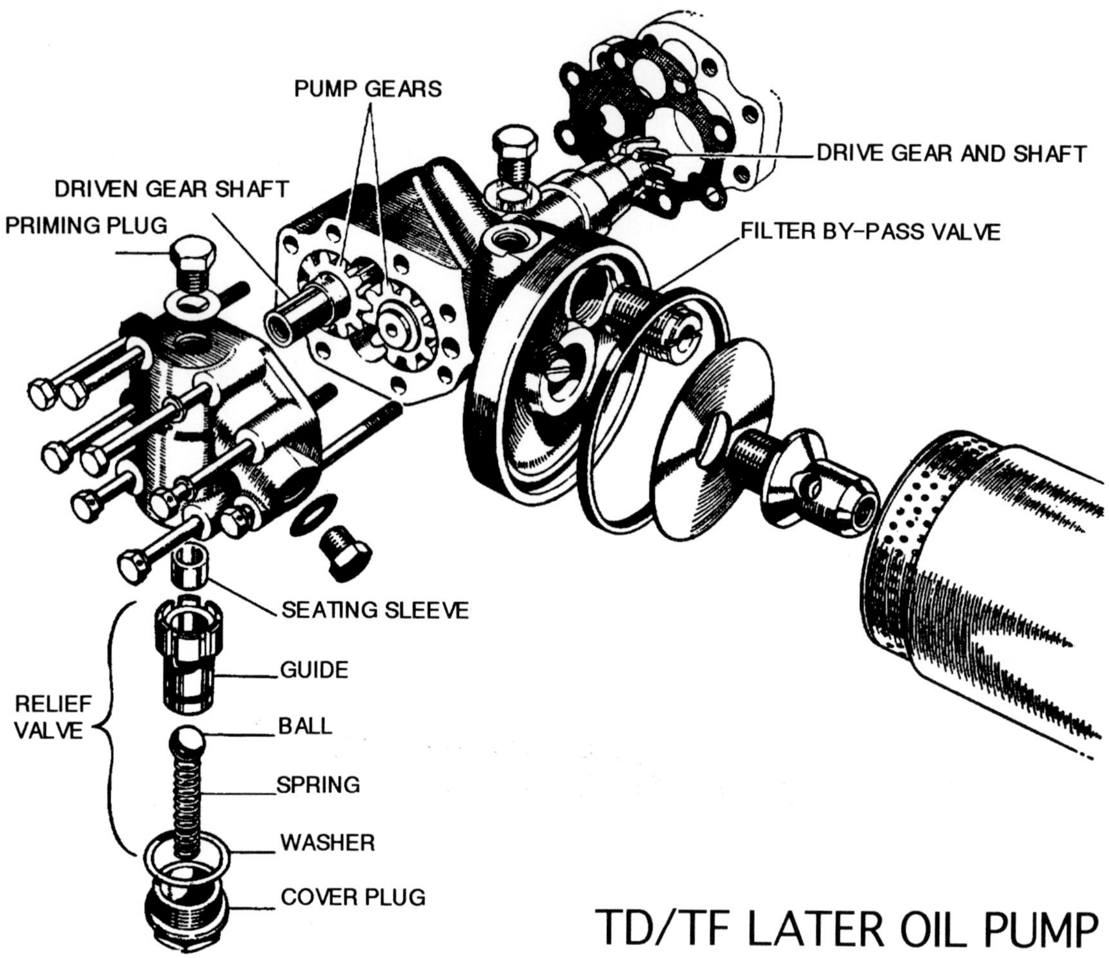

TD/TF LATER OIL PUMP

that it is free and that the gears are not stiff to turn. I usually fit this after the camshaft is in place. Take care tightening the bolts fixing the oil pump - they snap very easily. For this reason use new bolts with copper washers under the heads and lock them with wire.

FITTING A LATER PUMP TO AN EARLY ENGINE.

The oil pipe between the filter and engine block must be removed and the threaded hole in the block closed with a quarter inch BSP plug. The oil filter and filter bracket should be removed and holes in the block holding the bracket must be filled by fitting the bolts back in place with washers to seal against oil leaks.

Remove the pump from the block and extract the brass plug with the threaded hole in the centre. This plug is the by-pass valve and it can be removed by screwing in an 8 x 1mm bolt as an extractor. There is a steel ball, cup and spring behind this plug which must be removed.

It is important that all parts of this by-pass valve are removed as the later style pump feeds oil to the engine through this tapping. The later style pump still uses eight bolts but some of them are a different length to those used on the early pump. In view of their fragility I suggest that a new set of bolts is purchased. The gasket between the pump and block should be renewed. The oil filter element, the same type as used on the MGA and early MGB, should be replaced and the pump primed by introducing oil through the priming plug whilst turning the engine over on the starting handle.

Priming the oil pump on T-series engines sometimes causes difficulty. With the earlier pump, which did not have a priming plug, I have sometimes resorted to removing the oil pipe and feeding oil into the pump whilst turning the engine backwards to fill the pipe into the sump. Another trick is to pack the pump gears with "Vaseline", which is a petroleum product, as this will help provide suction when the gears turn. In any event the pump must be primed before starting the engine.

MODIFICATIONS TO INCREASE POWER.

The XPAG engine is capable of being tuned to produce considerably more power without sacrificing reliability unduly. Presuming the crank and rods have been crack tested and the bottom end and oil pump re-built, raising the power by 15 - 20% will add greatly to the the ability of the car to keep up with modern traffic, and pull a higher axle ratio, without any increased chance of breaking the engine.

The cylinder head changed during the production run of the engine and the later cars benefited from larger valves, fed by larger carburettors. The earlier head can be modified to take the larger valves and can be polished and gas flowed. It can also be skimmed to increase compression slightly. The standard head was 76.75 mm thick and reducing this to 74.37 mm increases the compression ratio to 8.6 : 1. The factory tuning booklet quotes an increase of about 10% for this modification, combined with polishing the head, ports and manifold, using the standard valves. If the valves are increased in size to those used on the TD Mark II then the power output would be further increased. The minimum thickness for the head was recommended as 73.575 mm, giving a ratio of 9.3 : 1

Modifications to the head are best left to specialists unless you have had some experience of the process. If the valve seats need replacing it may be wise to ask for hardened valve seat inserts so the engine can run on unleaded fuel. The time will come when leaded fuel

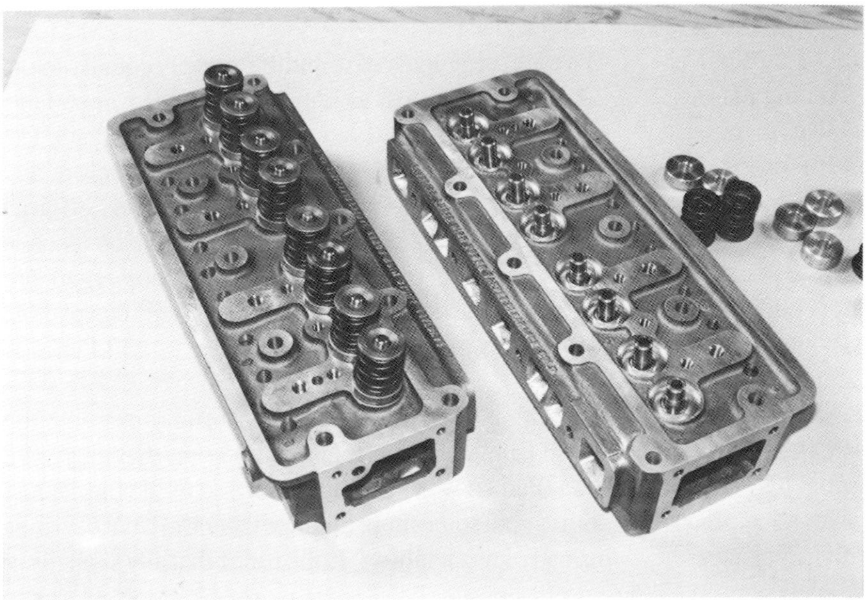

Aluminium cylinder heads, as re-manufactured by Brown & Gammons.

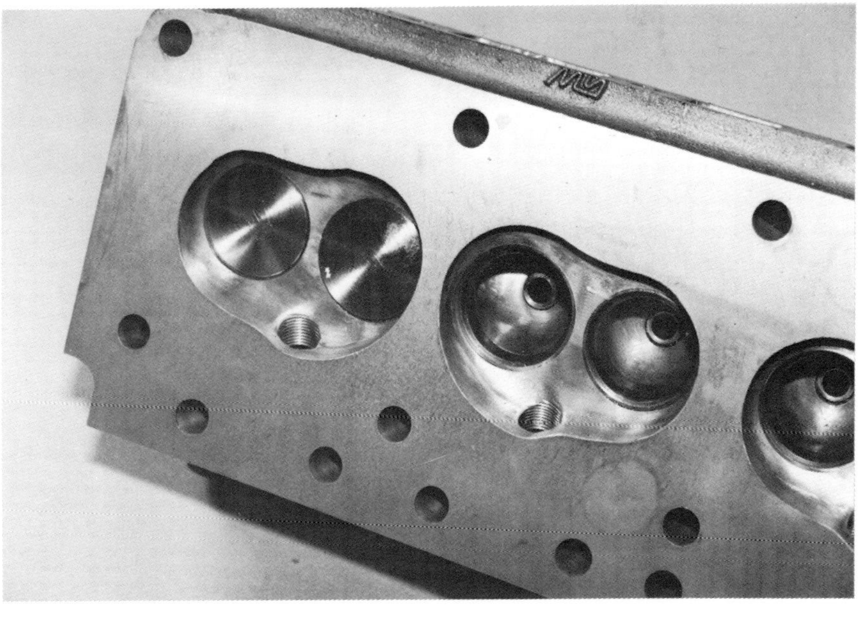

The over-sized valves fitted to this head.

will be very hard to find. As modern fuels do not seem to be getting any better it is probably wise not to aim for too high a compression ratio. When calculating the ratio remember that if the engine has been bored out to an oversize the ratio will increase unless the combustion chamber has been proportionally enlarged. Conversely if the chamber is slightly enlarged to accommodate the bigger valves the ratio will fall if the head has not been reduced in height. When measuring a head to determine if it has been skimmed remember that sometimes the top face will have also been machined to clean it up.

If the head has been reduced in height the valve gear will have to be modified to suit by shortening the push rods.

One of the premier suppliers of competition parts for the XPAG engine, Brown and Gammons of Baldock, have had re-manufactured the Derrington Aluminium cylinder head. This head was a favoured tuning aid in the 1950's and produced a considerable power increase over the standard item. It is now made with inserts to enable it to run on unleaded fuel and fitting this head and an extractor manifold will increase power by about 25%. The head is a little wider on the inlet side than the standard one and this brings the carburettors rather too close to the sides of the bonnet, especially on the TF. The engine can be off-set slightly by elongating the holes in the mounting brackets (which means the starting handle will not fit) or carburettors with shorter bodies fitted or the air cleaners omitted.

Brown and Gammons also supply special oversized and racing pistons as well as stronger big end bolts and cylinder head studs. The new camshafts they supply for road use have better profiling and this produces a small power advantage over the standard one.

Supercharging was often used to increase power output and the factory tuning book gave details of fitting a Shorrock supercharger kit. Al. Moss, the original founder of Moss Motors, has run his TC supercharged for many years and has also supercharged other XPAG engined M.G.s including a Y-type saloon. He advocates supercharging as a reliable way to increase power, especially at low revs, and it is appropriate that Moss Motors are now listing complete supercharger kits in their current catalogue.

THE TA ENGINE

The main principals of re-building apply equally to the TA unit and it is just as important that all components are carefully checked before re-assembly. Crack test rods and crankshaft as a matter of course before any further work is carried out on the engine. A large number of TA crankshafts will be found to be cracked although these can be repaired by submerged arc welding. Balancing the components will improve the smoothness of the engine.

In addition to these crack tests carefully examine the big end and main bearing caps for signs that they have been filed. It was a standard practice to file the caps of both to gain some more mileage from worn out bearings instead of re-metalling. Look to see that any such filing

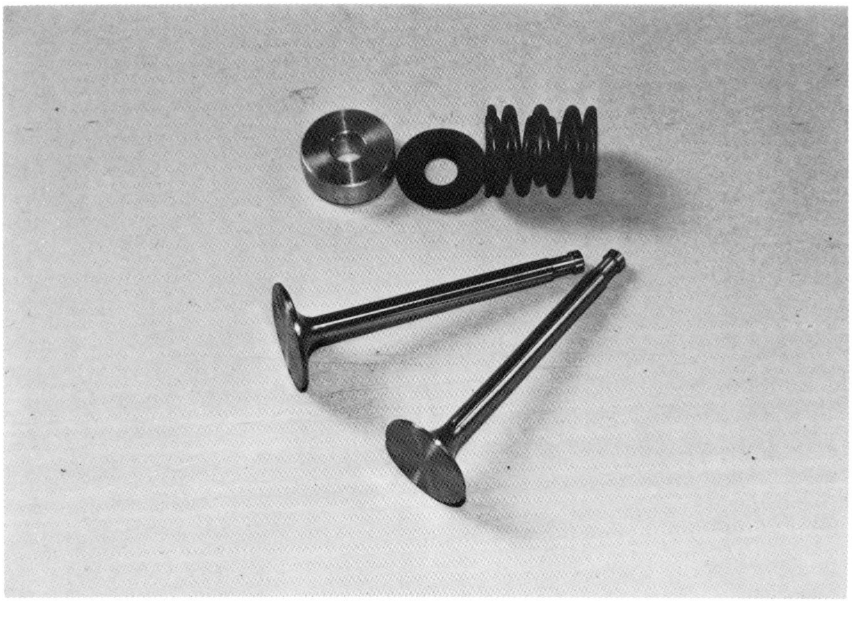

Modern shorter valve springs with an aluminium spacer for XPAG engine.

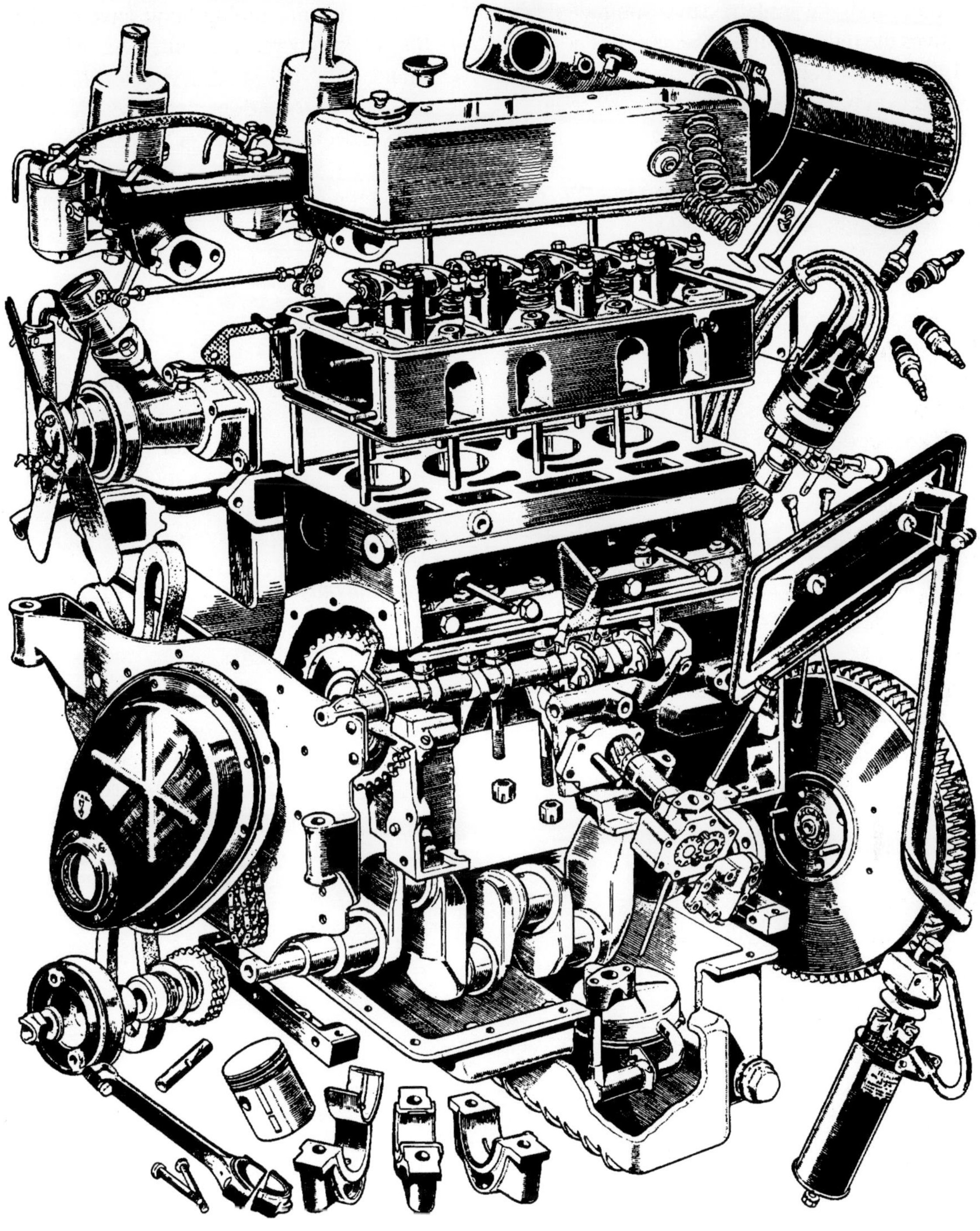

MPJG TA ENGINE

Two views of the TA engine bay. All TAs had chromed rocker covers but only the later cars were fitted with the group nipple lubrication system seen here.

has been done keeping the surfaces both flat and parallel to each other. If you assemble a con rod and cap with just one bolt and hold it against a light you will see if the surfaces line up correctly. If they do not they will have to be ground parallel before the bearings are re-metalled.

It used to be necessary to scrape newly metalled bearings when assembling an engine but modern machining techniques have made this practice redundant. Reject any re-metalling that needs scraping to achieve a good fit. If the main bearings are re-metalled they will need to be line bored to fit the size to which the crankshaft bearing surfaces have been ground. As we will see when assembling an engine with shell bearings, the crankshaft should be free enough to turn with gentle pressure at the flywheel - it must not be too tight. There is an oil hole in the conrod that allows a squirt of oil onto the cylinder wall with each revolution. After re-metalling this hole must be cleared of white metal.

As with the later engines, trial fit the con-rods on the crankshaft to ensure it will rotate freely and that there is sufficient end float - return to the machine shop if you are not satisfied.

One major problem with the TA engine is leakage of water through porous castings in the head and block. Foundry techniques in the 1930's were not very advanced and these castings in old age are often frail. It is now possible to repair porous castings with a new process often used currently on aluminium castings. This process involves subjecting the casting to a vacuum and then injecting a two-pack resin which is drawn into any flaws or porosity in it by the suction exerted by the vacuum. This process, Ultra Seal, has been used successfully on a number of porous blocks to my knowledge, by the company in Slough bearing that name and it is certainly better than having to use a radiator sealant as was the case in earlier years! Some oil leaks on TA engines occur because either the head or block faces are not flat so these should be checked.

In addition to being porous TA blocks are prone to crack if the head studs are over-tightened. The studs can pull out completely from the block. When tightening down the head it is best to start with a maximum of 25lb./ft. gently increasing this to around 35-40lb./ft.

The cork clutch can be relined by specialists, such as Sports and Vintage Motors, and should be soaked in engine oil for at least fortyeight hours before assembly.

As the clutch runs in oil there is a fair amount of this in the clutch housing and to avoid leaks great care should be taken to seal this on assembly. Use a gasket sealant as well as the gasket but be careful not to crack the clutch housing by over-tightening the bolts.

Some TAs run very hot and as the design of the water jacket in the head seems to be at fault. A few owners countered this by removing the thermostat after making sure that the radiator is working properly, re-coring if necessary. The water pump is a weak point on this engine and although kits are available to rebuild the original unit it appears that better results are obtained by fitting one of the completely new pumps currently available. If the old pump is rebuilt it is important to lap the carbon seal to the pump. As with the later cars it seems important not to run the engine with the ignition too far retarded. Modern fuels appear to need the ignition advanced three to four degrees.

As with the XPAG engine the side cover plate bolts should not be over-tightened as this can distort the plate and cause leaks. Using one of the "instant gaskets" will help seal the cover without needing to bolt down too tight.

TA AIR CLEANER

ENGINE ASSEMBLY

We will look at the assembly of an XPAG engine as demonstrated by Roger Thomas, who builds M.G. engines as part of his restoration business.

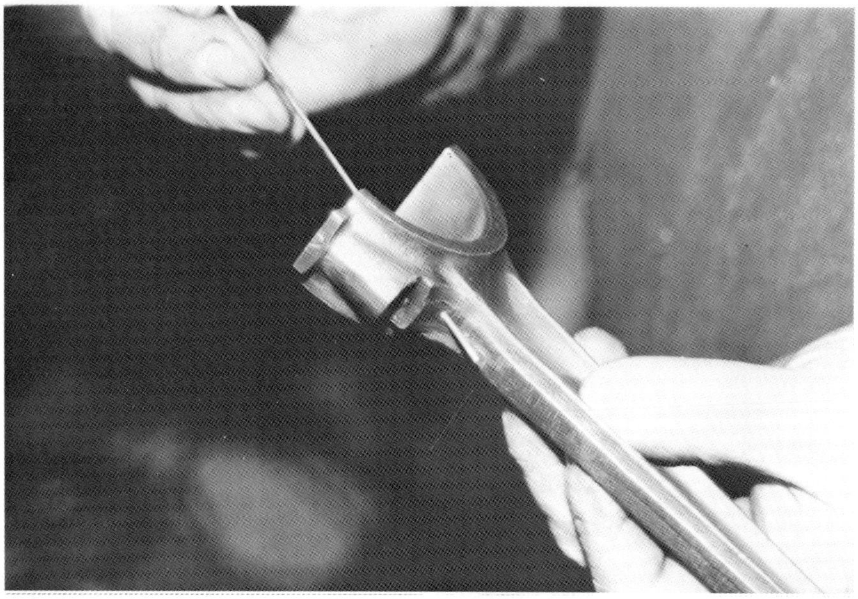

Before any assembly starts the fit of all bearings must be checked. Here Roger is cleaning out the oil hole in the con-rod and fitting a big end bearing. It is essential that the cap is fitted the right way round - there are marks on the cap and rod to help with this. When the rod is fitted to the piston the oil hole must point the right way. It is there to allow a jet of oil to squirt on the thrust face of the cylinder wall at each turn of the crank-shaft.

All oil drillings in the crankshaft must be clean and clear. Here, after cleaning, the last specks of dust are removed with the air line before the big end bearings are tried for fit on the journals. The bearings should be oiled and assembled with the old big end bolts - keep the new bolts and locking nuts for final assembly. Each bearing must be an easy fit on the crankshaft with a small amount of end float. If fitting a new crankshaft it may be necessary to relieve the edges of the bearings to give some end float. With the crankshaft turned horizontal the rods should fall freely under their own weight. If any of the bearings are tight check to make sure that the end cap has not been filed at some time. The flywheel to crankshaft bolts must be new and fitted with lock washers and wired.

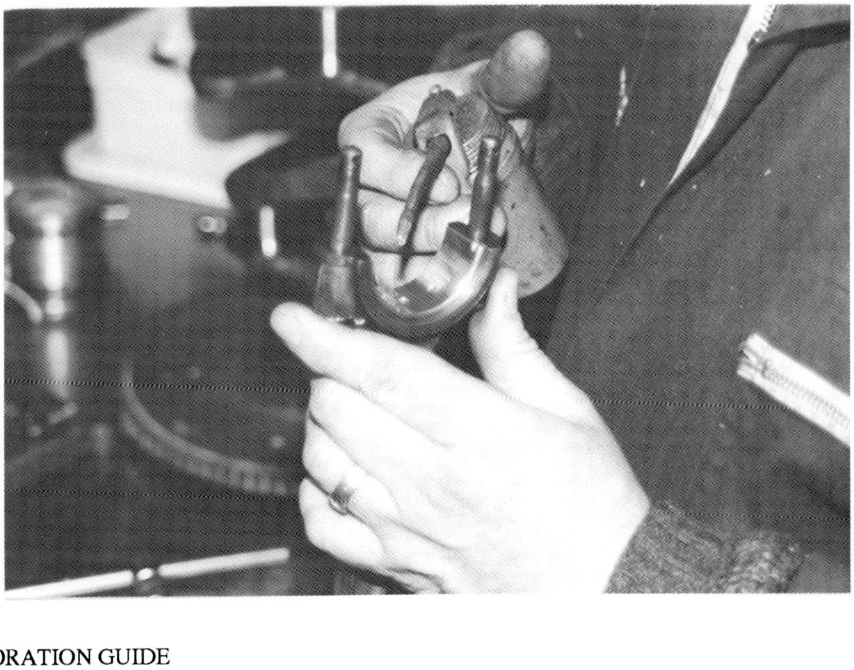

Before final assembly the oil pump must be trial fitted to the block if any new components are used. This can either be done first or, if it is to stay in place, after the camshaft is installed. The pump is fitted this early so it can be turned over easily to ensure the gears are not binding. Once the crankshaft and timing chain are in place it is not possible to judge if the pump gears are rotating freely.

Fitting the camshaft. The centre bearing splits in two and is fitted over the camshaft when it is in the block, before being eased into its housing. The oil holes must line up with the drillings in the block. The bearings are held in place with dowel bolts and the holes in the bearings that accept these must be aligned before the bolts are fitted.

When fitting the camshaft care should be taken not to dislodge the plug at the back of the engine, if this was not removed when stripping it down. It is vital that the plug is correctly fitted to prevent oil loss into the bell housing. The front securing plate seen here establishes the end float in the camshaft.

After checking the rings in the cylinder bores to ensure they have the correct expansion gap (unless this was done by the machinist) they are fitted to the pistons. The gaps in the rings are staggered round the circumference of the piston.

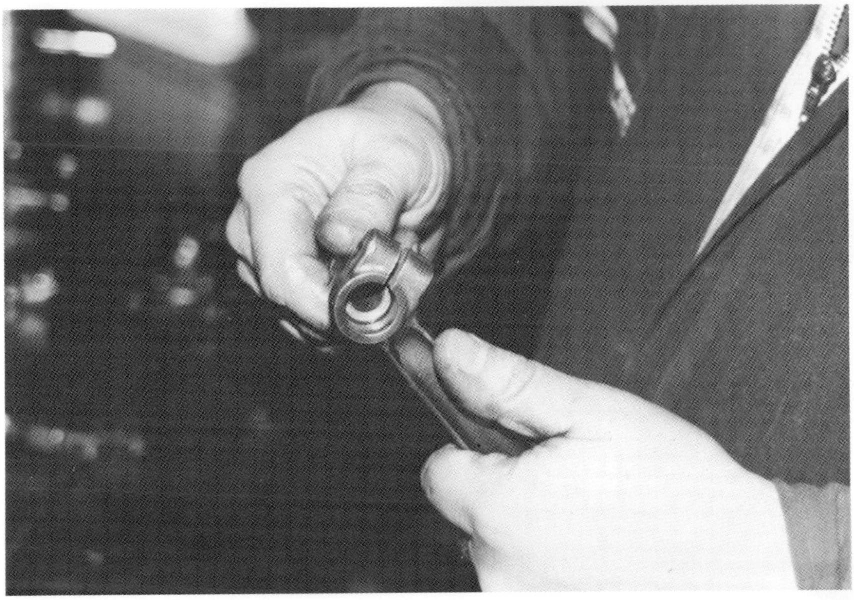

The gudgeon pins are trial fitted to the rods before being pushed into place in the piston. The piston should be fitted with the "front" mark on the crown in the correct relationship to the rod - the pinch bolt is on the right hand side of the engine.

Great care is needed to ensure that the pinch bolt is tightened correctly. Here a rod is passed through the gudgeon pin and held in the vice to support the piston while the bolt is tightened. With pistons having longer skirts the job is more difficult as a box spanner or socket is needed and neither will fit between the head of the bolt and the rod. The only answer is to cut a segment out of a box spanner just wide enough to enable the bolt to be tightened a turn at a time. The bolt should also be "Locktited"

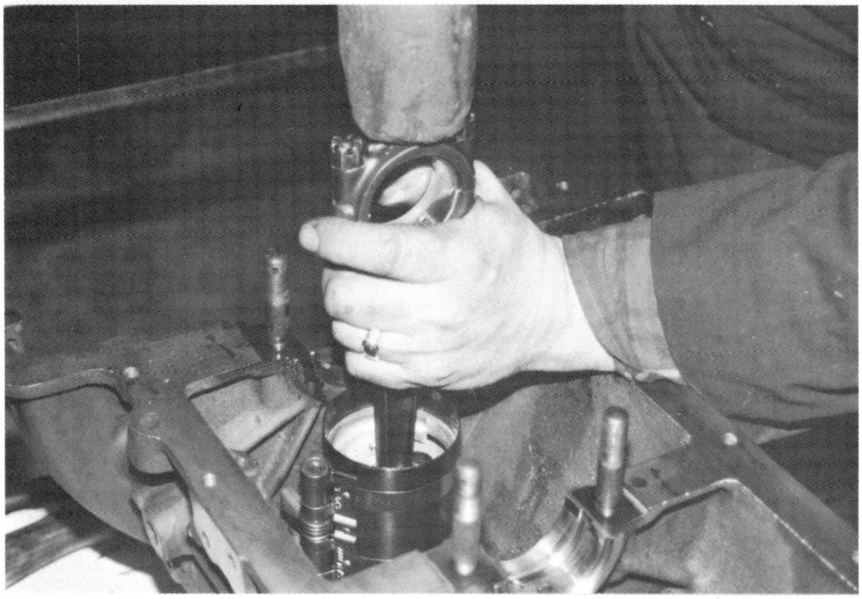

The pistons are installed from below and the crankshaft lowered into place. The big end caps and main bearing caps are fitted making sure that they are the right way round. As each bearing is tightened try the crankshaft to make certain it revolves freely. It should be easily turned with one hand at the flywheel. All bearings should be oiled when fitted or coated with an engine assembly compound such as "lubriplate". Finally all the nuts are tightened to the correct torque setting. It is probably wise to replace all load bearing studs and bolts when rebuilding as a failure could be expensive!

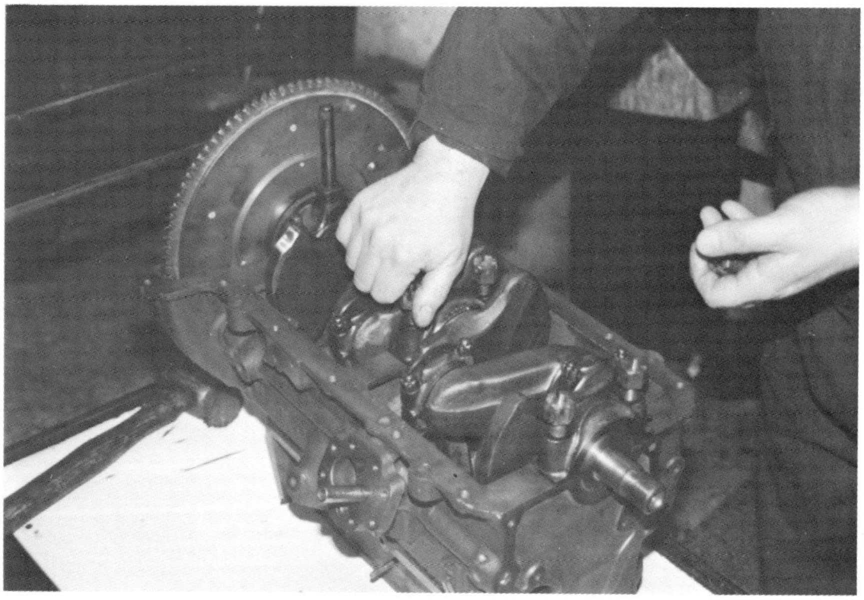

The clutch cover is bolted down using a bar to centralise the driven plate.

The timing gears are installed after the front engine plate is fitted. Here, as the timing chain has a removable link, the gears alone are being installed. If the chain cannot be opened out it must be fitted along with the gears. In the latter case I would fit the tensioner after the chain was in place. The tensioner should be sealed to the block, making sure the oil feed hole is clear, and the bolts wired in place.

The chain is fitted and joined. Care must be taken to make sure the timing marks on the gears are in the right place and that the links on the chain are counted and it is fitted as shown in the manual.

Rocker shafts. The old one on the right shows considerable wear because it was incorrectly installed. There is an oil drilling at one end that should line up with the drilling in the rear support bracket. If this does not line up the rocker arm bushes will be starved of oil and rapid wear will result. Note the screw that plugs the hole in the end of the shaft.

The support brackets, washers and shaft. There are two "D" shaped washers that fit in the front and rear brackets and two round washers that fit the centre brackets, at the same time locating in slots in the shaft.

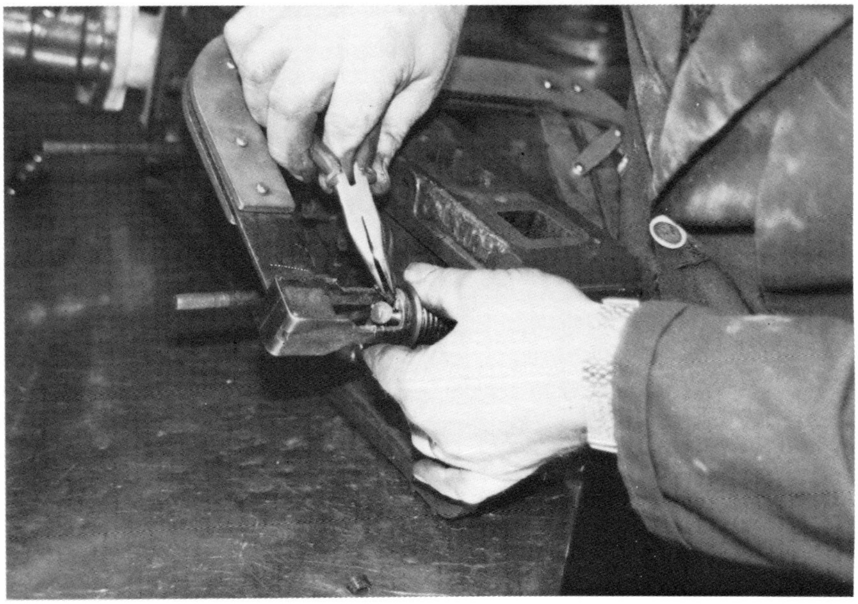

Fitting the valves. A dab of grease will hold the cotters in position until the valve spring compressor is released. The oil seal is located as shown on the next page - fitting this in the wrong place is a major cause of burning oil.

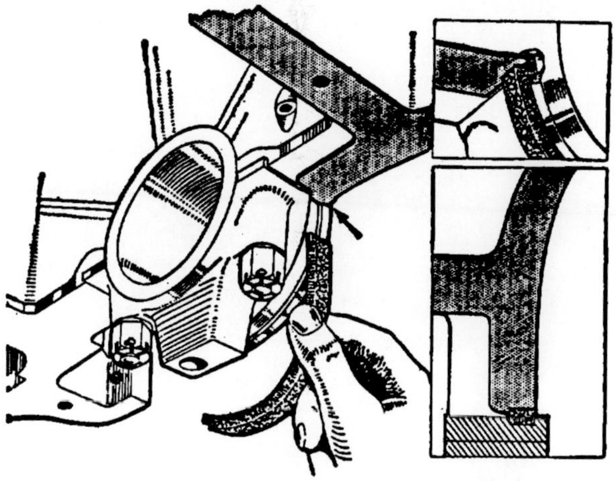

When fitting the cork seal for the rear main bearing it must have the stepped end over the end of the sump gasket.

FITTING THE SUMP

When fitting the timing chain cover and sump make sure the seals are placed correctly. The gaskets supplied are often not a good fit and will need adjustment. When installing the sump fit and tighten the two bolts at the back, those alongside where the gasket forks to join the bell housing, and then work forward fitting the rest of the bolts.

The correct fitting of the front oil seal and sump gasket is on the right and below is shown the valve stem oil seal assembly

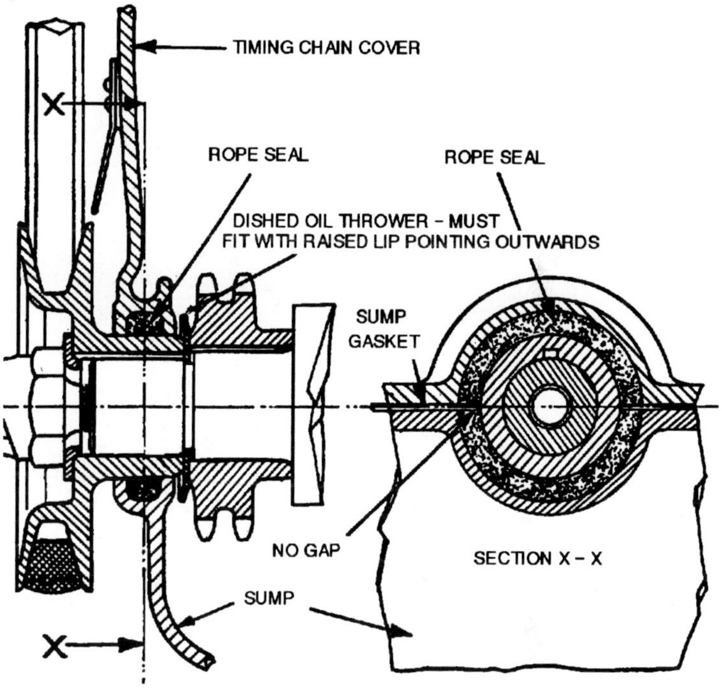

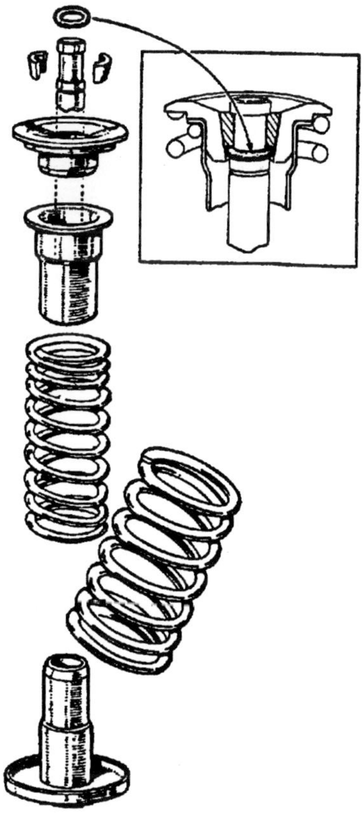

SUMP BOLTS - ALL 8X1 mm THREAD BUT HEADS DIFFER.

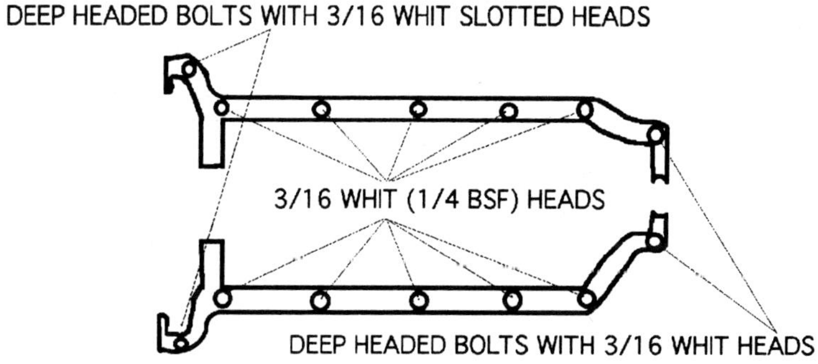

BOLTS FIXING BELL HOUSING TO BLOCK ALL 8X1 mm
WITH 1/4 WHIT (5/16 BSF) HEADS

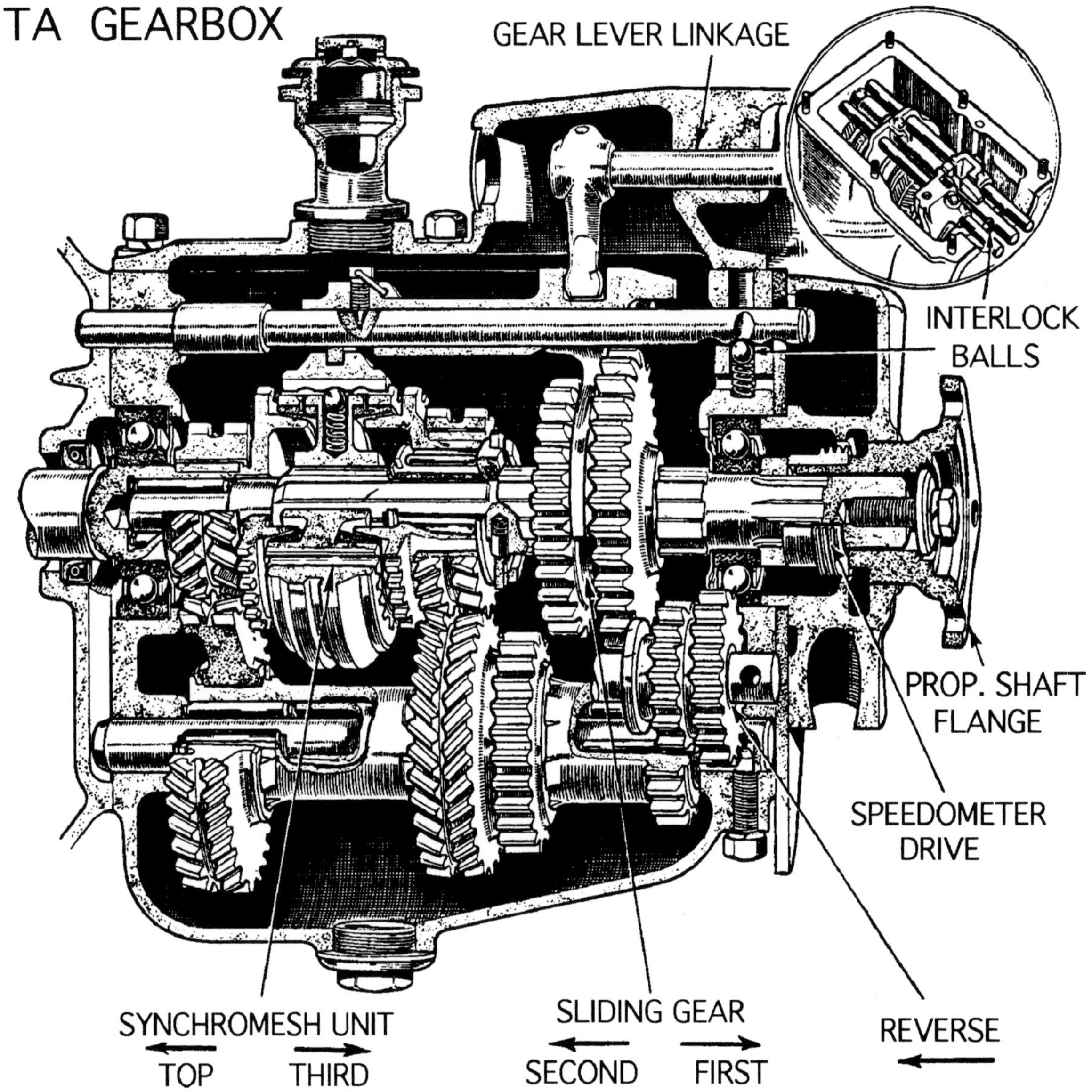
TA GEARBOX
GEAR LEVER LINKAGE
INTERLOCK
BALLS
PROP. SHAFT
FLANGE
SPEEDOMETER
DRIVE
SYNCHROMESH UNIT
TOP
THIRD
SLIDING GEAR
SECOND
FIRST
REVERSE

CHAPTER SIX

OTHER MECHANICAL COMPONENTS

As I said in the forward, it is not my intention to try to re-write the workshop manuals for these cars, merely to try to give additional information where I can. Quite full descriptions of gearbox repair in particular are given in the manuals and, as with the engine, the main difficulty is establishing the extent of wear and if parts need repair or replacement.

THE GEARBOX

The TC and earlier gearboxes are very strong units and capable of taking considerably more power than is produced by the standard engine. If you have driven the car you will know how the gearbox feels and if any faults are apparent. Anyone more used to driving modern vehicles may have found that the synchromesh felt weak and that it was all too easy to crash gears if changes were rushed. Take the age, and design, into consideration before deciding if it is faulty. Likewise expect some noise in the lower gears, although this should not be excessive. Jumping out of gear is sometimes a symptom of wear in the box but is more often a selector problem. However, early TD boxes did suffer from a tendency to jump out of third gear and later ones were modified to overcome this.

Over the years it is possible that previous owners may have treated your gearbox to a diet of incorrect oil, or even some patent "wonder goo", which may have affected the bearings and the operation of the synchromesh cones. If the cones are not badly worn but do not seem to be working very well then it may help to remove some of the polish from their surface. Use a light grinding compound and a little oil to gently lap the synchromesh surfaces together removing this polish. After doing this it is essential that ALL traces of compound are removed, especially from the little grooves, before re-assembling the box.

Almost all of the parts for the TD/TF gearboxes are currently available, including gears, selector forks etc. The TA/TC gearboxes are not quite so well served but bearings and oil seals are available but gears are not. If you find you need a part that is not presently produced you will have to search for a good secondhand item or have it made. At a pinch try looking at gearboxes from other Nuffield cars as many parts are common although the gear ratios used were often different. The bell housing can be cracked or badly worn round the clutch lever, particularly on TCs. There are now new housings available if yours is beyond aluminium welding.

If you are happy with your gearbox and do not wish to fully re-build it but would like to replace the front oil seal then this can be easily done without dismantling. The only problem is sliding the seal over the splines without damaging it and a piece of shim stock, or similar, over them before sliding the seal on will help.

After re-assembling a gearbox, or when looking one over, always make sure that the rear nut securing the prop shaft flange is really tight.

REAR AXLE.

The rear axle on the later TD/TF is totally different from the earlier unit so we will look at this later.

The TA, TB and TC axle is a simple one to work on although it has some weaknesses. As we saw when looking at the chassis the rear hubs have a tendency to leak oil on to the brake linings unless care is taken to assemble them carefully. I cannot stress enough the importance of making sure that the bearing is gripped by the rim of the hub and not allowed to move, and also that the small set screws are not placed in the wrong holes - they go through the UN-TAPPED holes in the hub, not those with a thread. There have been cars where the rear axle casing itself has been bent slightly in an earlier accident causing problems for a subsequent owner unaware of the real cause of all his problems! The mounting plates on the axle casing do sometimes split and if this has happened reinforcing pieces welded to the bracket will effect a cure.

Two different axle ratios were fitted to the earlier cars, the TA had 4.875:1 and the TB and TC were fitted with 5.125:1 gearing. Owners of the later cars often fit the earlier ratio to reduce the engine revs. at any given road speed which is a great advantage on modern roads. There is no difficulty fitting the alternative ratio as long as the backlash is correctly set up as described in the manual. The gears, bearings and shims are available at the time of writing.

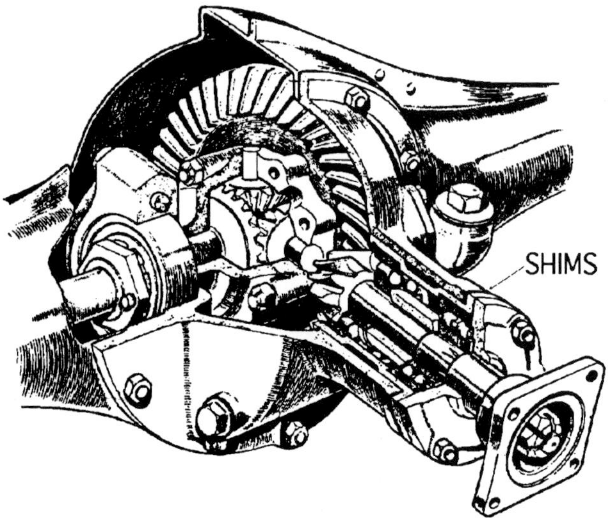

TA, TB, & TC DIFFERENTIAL

The original arrangement of oil sealing for the pinion left a lot to be desired as there was no proper seal fitted just a scroll to try to slow down the flow of oil! It is now possible to fit a new pinion cap which incorporates a modern lip seal. When re-filling the axle remember not to over fill it. DO NOT fill to the top of the filler plug hole - remove the level plug and only fill up until oil flows out of the level plug hole. Over-filling is probably the main reason for leaks.

The half-shafts on all the T-series cars can fail after a while, particularly if the engine is tuned or the car used for competition. Never swap over shafts when fitting new hubs as they can fail once the stress is reversed. A modification used on some competition cars is to substitute a Ford Escort Mk 1 differential which will fit in the housing once the bolt holes are re-drilled. Part of the rim needs a small amount of metal removed. New half-shafts are required but these are available from a number of suppliers as well as from the M.G. Car Club T Register Spares Secretary and usually come with the hubs installed. The rear prop-shaft flange will need changing but this can be done at the local prop-shaft specialist. A number of different ratios were fitted to various versions of the Escort which now gives some choice. The only difficulty is that even supplies of old Escorts are drying up!

The TD/TF axle is a much more difficult unit to work on and quite a few special tools are needed to set it up if new gears are fitted. This is a pity as you do really need to raise the rear axle ratio if you wish to travel far on modern motorways at any reasonably rapid pace.

Instructions in the manual are good but presume that you have access to all the special tools to do the job. A few years ago the Southern California Vintage M.G. Club produced a very comprehensive manual on changing ratios, including drawings of home made substitute special tools. This booklet, available through Moss Motors, also contains details of speedometer gearing changes.

The highest ratio originally available for the TD/TF was 4.55:1 and sets of these gears are at present available. The higher 4.3:1 ratio fitted to MGA 1500 and 1600 models (but not the Mark ll) as well as Morris Oxford and Austin Cambridge saloons gives even more relaxed cruising at the expense of top gear performance. Fitting a 4.3:1 gear set increases the miles per hour at 1,000 r.p.m. to 17.16 from the 14.4 of the standard TD 5.125:1 ratio. This means that the comfortable engine speed of 4,000 r.p.m. carries you along at over 68 m.p.h. against just under 58 m.p.h. with the lower ratio. As T-type engines wear rapidly at high revolutions the change would appear well worth the effort involved.

THE BRAKING SYSTEM.

The braking system on all T-types is simple to understand and work on and good details are given in the manuals. The system for the TA, TB and TC is very different to that on the later cars and there are no major common parts between the two systems.

The earlier system is really a development of the cable brake set up of the O.H.C. cars. The handbrake cross-shaft and the handbrake cables are fitted in a similar way to the pattern set for all models from the C-type to the PB. Restoration of the handbrake system is usually confined to cleaning and repainting the various parts and checking that the bearings in the brackets supporting the cross-shaft are free and well lubricated. The handbrake lever was chromed before assembly and if this requires replating you will need to dismantle the whole unit. Often restorers will just paint the lever black and re-chrome only the knob - telling themselves that not a lot will show once the gaiter is in place!

New brake cables are available but if your car still has the original cables with the brass fittings it is better to have these re-made with new outer and inner cables as the new items have aluminium fittings which do not look quite as good. When adjusting the cables set them so that the brakes start to operate with the levers in the vertical position. Remember to replace the rubber bump stop if this is missing. The cables were originally lubricated either from the grouped system on late TAs and TBs or from two grease nipples fitted to brackets just in front of the rear wheel. Thcsc nipples and short pipe runs are often missing from restored cars but replacements are available from suppliers like Moss and Abingdon Spares.

The pipe runs for the hydraulic system on the earlier cars is in a larger bore piping than the TD/TF but sets of these pipes are widely available. Correctly the pipes are shielded with a steel spring sheath where they are exposed to the risk of damage and stainless steel coil spring is available for this purpose and should be fitted. The master cylinder and slave cylinders are no longer available from the original suppliers so either the original units will have to be re-furbished or new replica items purchased. In the United States it has been possible for quite a few years to have the old cylinders re-sleeved with brass or bronze sleeves and these are then fitted with new rubbers as they are easily available. Although I have no experience of this service myself I know of quite a few people very happy with the results they have had from re-built units. I have purchased new replica parts for a recent TC re-build; these were very good and with silicon brake fluid should give many years service. However, there have been some problems with replica master cylinders and these have mainly centred around pressure build-up in the system.

As a car is driven the brake fluid is heated and expands. Usually the excess fluid drains back into the master cylinder reservoir through a small hole in the bore of the master cylinder, exposed when the piston is fully retracted. On some cylinders it seems that the rubbers in the piston can cover this hole, stopping the expanding fluid returning to the reservoir. As a result the build-up of pressure will apply the brakes, usually at one wheel first, creating more heat which applies the brakes even harder. The first reaction of the owner is to blame the slave cylinder or brake adjustment rather than the real culprit - the master cylinder. Of course the usual reason is that the brake pedal has been wrongly adjusted

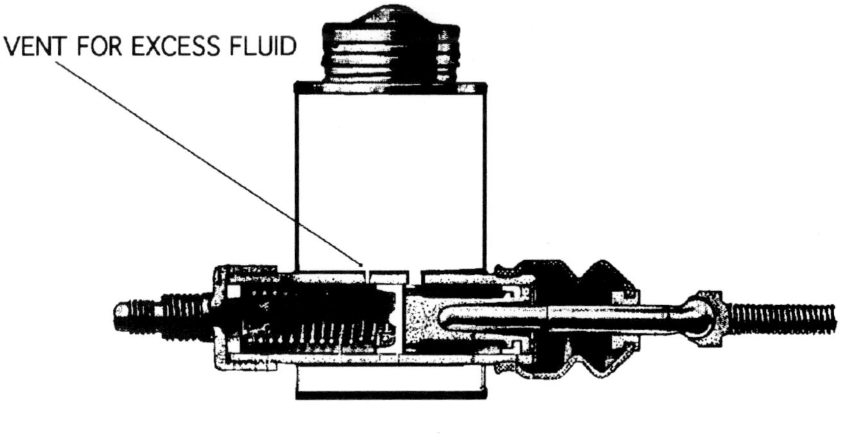

TA, TB, AND TC MASTER CYLINDER

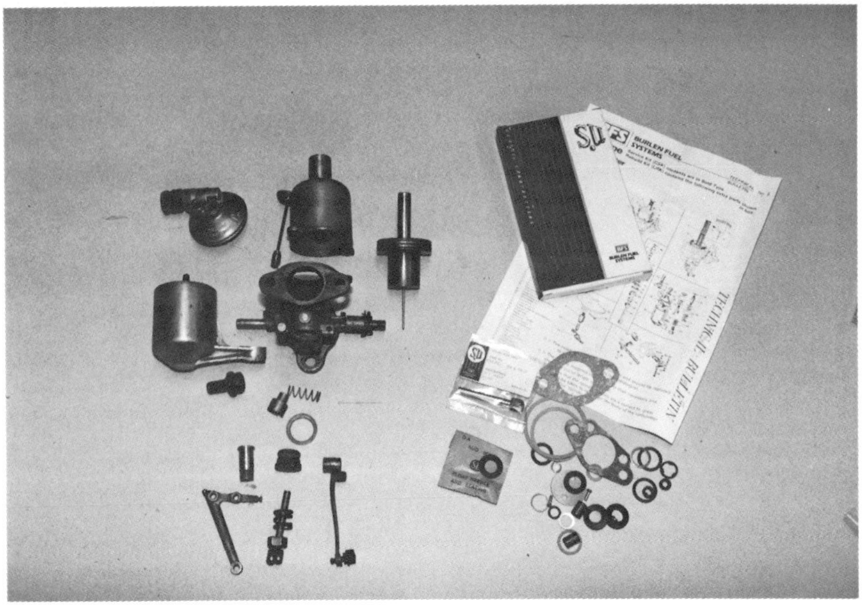

Kits are available to re-build all S.U. carburettors.

Home-made heat shield fitted to a TC. Note the bracing bracket at the front.

Another home-made heat shield fitted this time to a TD Mk ll. Only one fold was necessary for this shield to fit well. Both of these can be removed quickly without disturbing the carburettors.

without sufficient free play. The piston must be allowed to fully return to clear the small hole.

I have had seen this problem on a couple of TC replica cylinders. On the first the bore of the cylinder was incorrectly machined, giving trouble from the start. On the other the car ran well for a year or so before the symptoms appeared and it seems the rubber seal had become softer during this period (one theory advanced implicated the silicon brake fluid although I cannot find evidence to support this) and the hole was now being covered. Rebuilding with new seals only effected a temporary repair and eventually an original cylinder was re-furbished and installed. Even on good original cylinders assembling the parts of the piston incorrectly can cause the same problem. If a sticking brake is evident, and you suspect the master cylinder, try opening a bleed screw slightly at one of the wheel cylinders. If the system is pressurised a small jet of fluid will come out of the bleed nipple and the wheel will free off.

The master cylinder is shielded from exhaust heat by a metal box incorporating a piece of asbestos as insulation. This box is often missing but replacements are available and it is well worth fitting one.

If the drums are either scored or badly distorted then good braking will be impossible to achieve. If the drums are only lightly scored then it may be possible to have them skimmed but this will need to be done very carefully if they are not to be ruined. Some replacement brake linings are available, on an exchange basis, bonded to the shoe but most come as separate linings which you rivet in place. This is not difficult to do but great care must be taken to ensure that there is no gap between the lining and the shoe and that the rivets are secure.

The later TD/TF system uses cylinders common to a number of other British cars of the period and, at the time of writing, Lockheed parts are available and it is possible to fit new wheel and master cylinders all round at a reasonable cost. These cylinders often turn up at autojumbles at low prices as well as being available from the major suppliers. Sets of the brake pipes are available although these are usually made of copper and I prefer to make up my own pipes in cupro-nickel as the colour of this looks closer to the original steel pipe. Having said that there is not a lot of the pipe work visible on the car when it is finished! I used one of the pipe flaring tools available to the restorer and recovered the cost of this from the savings made on my first set of pipes. When installing the pipes make sure that they cannot be rubbed by the rear axle check straps or by the damper arms. This is particularly important on the TD Mk ll and the factory then supplied a longer pipe for the rear brakes on this model to overcome the problem.

Unlike the earlier cars it is quite easy to fully dismantle the handbrake lever for re-chroming and this is well worthwhile as the lever is very prominent in the cockpit. There were no extensions fitted to the grease nipples on the handbrake cables of the later cars.

CARBURETTORS AND HEAT SHIELDS.

The carburettors changed from model to model but the basic design and the restoration techniques used are much the same on all of them. Repair kits are available and some of these contain good basic instructions. Wear is usually limited to the throttle spindle, the jets and the needles and these are all easily repairable. Some carburettor bodies, especially on the earlier cars, suffer from distortion after a time. A TC I rebuilt had a pair of carburettor bodies with the flanges so distorted that even the thickest gasket would not have prevented air leaking in. Many a TC owner has wondered where the high pitched scream he can hear at low revs is coming from - it sounds dreadful and visions of some serious bearing problem can rush to mind! The body can also distort enough to cause the butterfly to jam in the throat and if this is the case replacement is the only answer. If the flange distortion is not too bad then careful re-facing will cure the problem but otherwise it will have to be replacement bodies.

The spindle bearings in the body can be re-bushed to take a standard spindle or reamed out to take an oversized one. Kits and reamers are available from Moss and other suppliers. Burlen Services in Salisbury offer a full re-building service as well as all the necessary parts. Care must be taken when reaming that the holes are reamed in line.

Once stripped, the carburettor parts can be cleaned using a proprietary carburettor cleaner but this MUST be used in the open air and you should wear rubber gloves and avoid breathing the fumes. All the various springs are available as replacement parts but these sometimes need some adjustment of the length to work properly. I would not use the plastic type of float chamber valve now supplied. This seems to make fuel

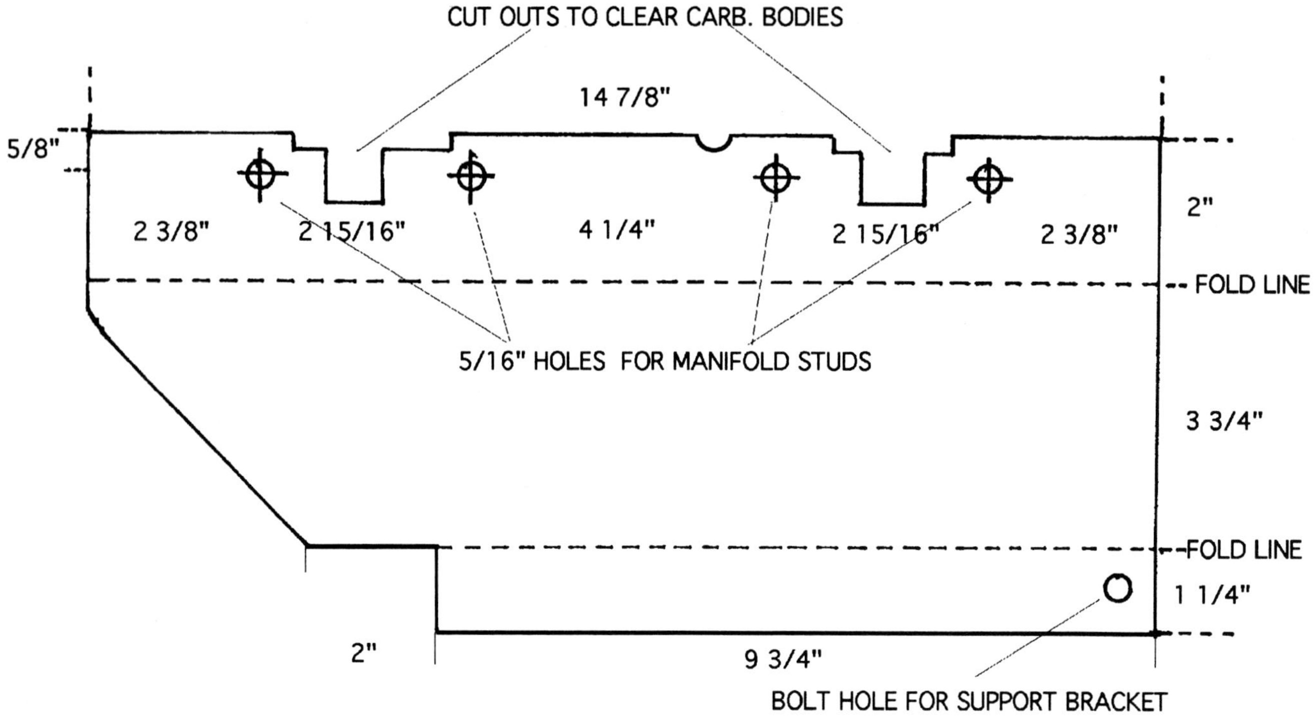

Drawing for heat shield for a TC. Some adjustment may be necessary for individual cars but shields made to the above pattern have fitted at least three different TCs without any problems. It is necessary, however, to ensure that the shield cannot foul the throttle linkage or springs at any point or they may jam open! The same basic design can be used for other models, amended where needed.

starvation problems worse when the engine is re-started hot. The heat from the exhaust manifold can cause the fuel to boil in the float chamber and perhaps this increase in pressure forces the valve closed? Anyway I have had some success by fitting the ball type "Grose Jet" valves in conjunction with a heat shield - which we will look at later.

It is essential that the three cork seals are soaked in oil for 24 hours before assembly. This not only helps them to fit without damage but also helps to make them leak proof. It is also vital that the jet is centred correctly. Spend as much time as you need to ensure that the piston will fall cleanly on to the bridge of the jet with a definite "clonk". If this is not done the car will never pick up smoothly from rest and the idling will be erratic.

When the carburettors are back on the car make sure that the mixture enrichment (choke) works evenly on both carburettors and that both jets fully return when the knob is pushed to the closed position. This will not be able to be verified until there is petrol in the carburettors but when there is make sure that all is well. It may be necessary to increase the strength of the return springs by slightly shortening them. Also check that the throttle levers are synchronised - this can be done roughly before the engine runs and later with a balancing tool. The overflow tubes on the float chambers must be free and make sure washers with cut outs are fitted, to allow the vapour or excess fuel to pass to the tubes.

Set the float chamber fuel level so that it is about 6 mm below the bridge. This is not an exact setting - just make sure that both are about the same height. Make sure that the fork is only bent where indicated in the manual. The height of "Grose Jets" are adjusted with washers.

As I have mentioned I have found heat shields an advantage in this country. In very hot climates more drastic measures are called for occasionally and some cars have been fitted with plastic fans to improve air flow and lagging for fuel lines and float chambers to cut down on the amount of heat absorbed from the exhaust manifold. The simple shield I have fitted is illustrated here and is based on a design that appeared in the T Register notes in "Safety Fast" a few years ago. I believe that this was based on a photograph of a 1950's competition car engine. The shield can be made of steel sheet painted silver or, better still, from stainless steel. The advantage of this design is that it can easily be fitted or removed without disturbing the carburettors. Another style of shield is marketed by Brown and Gammons and this fits between the carburettors and the inlet manifold. It looks better than my shield and gives a more "factory fitted" appearance but at the expense of not being so easily installed or removed.

There is no doubt that if some sort of insulated blocks could be fitted between the carburettors and the shield this would also help cut down heat transfer. The only difficulty is the lack of room to extend the carburettors outwards without hitting the bonnet sides and they might interfere with fitting the standard air cleaner.

The air inlet pipes on the TA, TB, TC and TD are left as bare aluminium. This was not polished to a high gloss by the factory although some owners now like to do this. The air cleaners were painted gloss black and a good powder coating firm can make them look very smart although to be original they should just be cellulose sprayed. The transfer, advising washing and re-oiling, that appears on TC canisters is available. The air inlet pipes and air cleaner on all but the earliest TD Mark lls differs from the standard item and is unique to this model. For the other models the trade can supply reproduction items.

EXHAUST SYSTEMS.

The exhaust system on the TA, TB and TC is made up of three sections. The front section running from the exhaust manifold to the silencer incorporates a piece of flexible pipe to absorb the engine movements. There is a bracket fixed to this pipe that is bolted to one of the bell housing bolts to prevent any twisting of the manifold joint. The silencer is bolted by four bolts to the chassis rail and the rear section of pipe is clipped to the rear of the chassis using a separate bracket.

The TD/TF system is also made up of three sections but at no point is this bolted directly to the chassis. The front section is fitted to a bracket attached by two bolts to the side of the gearbox casing. This bracket supports the front of the system, relieving stress on the manifold joint, but is often missing from cars. The silencer and rear pipe are held by two brackets, with rubber interleaving, to the two rearmost cross tubes.

On a car used infrequently there is an advantage in fitting one of the currently available stainless steel systems. Unfortunately some fit better than others and it may be necessary to adjust the angles of some of the bends in the pipes to get the system to fit the car. Although made of large diameter pipes I have found it possible to adjust bends quite large amounts using hydraulic jacks between the pipe and the chassis, making sure that any bending strain is not transferred to the exhaust manifold.

WHEELS AND TYRES.

If the car is fitted with wire wheels it is quite likely that these will need some attention before they can be painted and re-fitted. At present new wheels are

The heat shield supplied by Brown & Gammons fitted to a TF. This type of shield is placed between the carburettors and the inlet manifold.

available if the rims and centres of the old wheels are unfit for further use but if these are serviceable there are a number of specialists that can re-spoke and true up the old ones. All the paint will have to be removed by sand blasting or chemical dipping before repair and either powder coating or stove enamelling makes a good final finish. The factory never supplied chromed wire wheels for these cars and I think that they look better just painted. All the T-series cars had silver painted wheels.

The disc wheels cause less trouble and seldom need more than a strip and re-paint. The hub caps were chromed steel although some of the replacement items seen recently are polished aluminium. The medallion in the centre had the lettering highlighted with red paint on the TD but the hub cap badge on the later Mark IIs and the TF was changed to black and white.

The TA, TB, and TC wheels were fitted with 4.50 x 19 tyres and these and the tubes are very easy to fit at home with just a couple of levers. "Swarfega" or soap can be used to lubricate the bead. The later cars were originally fitted with cross-ply tyres but modern radials will improve the roadholding and are easy to obtain. If there is a choice of tread patterns one without a too "modern" appearance is best. The wheels will need to have tubes fitted.

CHAPTER SEVEN

ELECTRICAL EQUIPMENT

In this section I will look at all the electrical components giving, wherever possible, details of the changes from car to car and also the original part numbers to help searches at autojumbles and flea markets. Before looking at the rest of the car I will deal with the instruments and other dashboard fittings. For reasons I will explain later you should not leave this part of the restoration until last.

THE INSTRUMENTS.

The speedometer and tachometer on the TA, TB, TC and early TD are chronometric instruments and the later TD and the TF are magnetic. It is very easy to see which type your car has as the early instruments have flat faces and the later have a dished rim to the dial. The chronometric instruments are reckoned to be very accurate and were much prized by earlier generations. However, I have not found them to be all that good trying to match odometer readings when following a rally route! With either type the revolution counter has a weakness - the gearbox on the dynamo and the cable drive. One reason for the early failure of the gearbox is the over-tightening of the securing nut and cracking the casing. Replacement gearboxes I have seen are often too stiff in operation. Original factory replacements were much freer, although I do not know why. Another problem is over oiling in use. The grease nipple fitted to the box seldom needs touching, otherwise grease will work its way up the cable to the instrument. If the gearbox does move it can cause a misfire by shorting out the distributor L.T. lead. I have seen one very cunning conversion of the tachometer to electronic with the gear box and cable only there for show - but I would not go that far myself.

If you are going to have your instruments rebuilt then the first thing to say is do not take them apart yourself. Talking to one of the leading restorers of these instruments, John Marks of Vintage Restorations, I was told that one of the major causes of difficulties for them, and cost to the customer, is repairing damage caused by over enthusiastic owners. The delicate lugs on the rims are bent and broken, cases split and pointers damaged. The good news is that almost anything can be replaced or repaired. Indeed a small number of replica instruments containing original chronometric movements are made. Luckily the manufacturers used similar cases for most of the speedometers they built. This means that an autojumble can often yield up an original instrument that may have completely the wrong dial but the same case and mechanism as the one you need for your car. The instruments fitted to M.G.s from the later J2 right up to the early TD were very similar, even though the dials differed. The only problem you will encounter is the fitting for the 30 m.p.h. warning light. The electrical contacts and external fitting were not present on all speedometers although, unless you really feel the need

The later style of TD instruments had dished dials.

for this warning light to work, you need not worry as the lack of this will not be visible.

The next warning I would give is to take your instruments to the restorer as early in the re-build as possible. He may take quite a long while to complete the job, especially if you are missing any parts of the instrument, and it is very frustrating to be held up at the end waiting for the instruments. When you give or send your instruments for restoration ask for a written receipt and price quotation and also indicate when you would expect to need them back. This will help you plan to meet the cost and also help the restorer get them back to you in time.

The clock fitted to the tachometer is a separate unit and this is often missing on reproduction instruments. On the TA and TB this was mechanical but the later cars had an electric clock. Remember that restoration of the clock is a separate job and will cost extra, over and above the cost of restoring the main instrument. Vintage Restorations will fit new dials if required to the TA, TB, TC and TD chronometric and late TD instruments but at the time of writing dials for the TF instruments are not available so it is essential to locate good original units for re-building. Those fitted to the Y-type were similar but not the same.

One thing to consider before sending the speedometer for re-building is the final drive ratio fitted to your car. If you have changed the specification then all you will need to do is to let the restorer have details of the number of turns per mile the speedometer cable makes.

There is a simple way for you to work this out as long as the car is all in one piece and capable of being pushed along on its wheels; disconnect the cable from the speedometer and make a small card pointer to fit on the inner shaft. Make sure the tyres are at correct pressure and measure the radius from the centre of the rear hub to the ground. Make a chalk mark on a rear tyre and push the car along in a straight line until the wheel has revolved exactly six times, counting the number of turns the shaft makes. Calculate the number of turns per mile by multiplying the number of turns of the shaft by 1,680 and dividing by the radius in inches. For example if the shaft turns 9.2 times multiply this by 1,680 and divide by the radius of 12.5 in. to make a total T.P.M. in this case of 1,236.

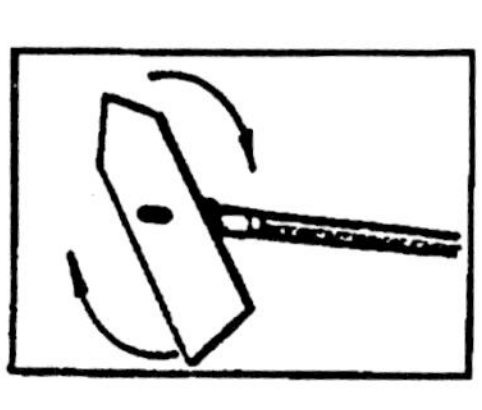

The dashboard layout of the TF allowed the factory to produce either L.H.D. or R.H.D. versions with little alteration.

Whilst thinking about ratios it may be a help to mention the size of the dynamo pulley. If you find your tachometer is inaccurate check the pulley to make sure it is the correct one for the car. If it is too small the tachometer will over-read as the dynamo will be turning over faster than usual. The opposite can also be true if the pulley on the engine is wrong. You will probably not need to worry too much about this as long as you have the correct engine speeds in mind when driving hard!

The smaller instruments fitted to the TA, TB, TC and TD are easy to find. As with the larger instruments, the cases did not change for many years and old units with black dials can easily be found and re-painted to the T-type style. The TC was not fitted with a temperature gauge as standard but many had them fitted as an optional extra. They were usually then mounted on the right hand side of the dash board, outboard of the tachometer. The early TD, likewise, lacked a temperature gauge but a dual water/oil pressure gauge was fitted from TD 13914. As I preferred the look of the dashboard without any extra dials I have fitted the dual gauge to a TC but had the lettering and pointer done in brown (the TD dual gauge had black lettering and pointer) to match the other TC instruments. The fittings to the radiator were also changed to the TC type. I was pleased with the result and did not get any adverse comments at meetings. Although the TD did not have the dual gauge fitted to early cars it was available as an optional extra soon after production started so you can feel free to fit one to almost any TD. The pre-war cars would really need to have a separate temperature gauge as I feel a duel unit would look wrong. A popular accessory then was the radiator cap thermometer and these can sometimes be found at autojumbles.

The ammeter changed from 20 amp. to 30 amp. when the TD instruments changed to magnetic. There was also a high beam warning light fitted to the speedometer from TD 17548, but you will now often find these on earlier cars as all the replacement magnetic type speedometers seem to have been fitted with the light.

SWITCHES.

LUCAS HORN AND DIPPER SWITCH 17A L15. Part No. 380191. This switch was used on the TA, TB,

TC and TD up to TD 18882 when a separate foot-operated dip switch was fitted. Correctly the switch sits on a flat, chromium plated, bezel and the black enamelled switch carries white lettering "D" and "H" to indicate dip or head lamps. Many replacement switches do not have this lettering. After the separate dip switch was fitted the combined unit was replaced by a horn push of identical pattern but lacking the switch lever and lettering. A service sheet was issued detailing the conversion of earlier cars and I expect quite a few were modified.

IGNITION AND LIGHTING SWITCH PLC. On TA/TB units there were four positions: Low and High charge, Side and Head lights. The lettering was recessed and filled with white paint. Post war, when a voltage regulator was fitted, the switch was Lucas PLC6 L114. Part No. 34018A. Fitted to all TCs and TDs. The correct type has "OFF S H" in white lettering at the top and LUCAS in white below. The switch fits in the metal centre panel with a wire clip that sits in slots cut in the side of the switch body. These slots are at various places to accommodate different thickness panels for other vehicles. There are quite a number of variations that may have been fitted over the years. Some later types had a window with the lettering appearing as the switch was moved and these have often been fitted as replacements.

THE CENTRE PANEL. There was just one basic design of centre panel fitted to all the T-type cars from the TA to the TD, although there were quite a few differences in layout of the instruments and switches between models and to accommodate export variations. On all the panels the raised rim was chrome plated and polished, whilst the centre was painted. The panel on the TA, TB and TC, up to approximately TC 5500, was painted black with white lettering printed on to identify the various knobs and switches. From then on the colour was changed to a metallic tan with black lettering although the lettering was omitted from the EX-U version. The TD panels were all painted the tan colour without lettering.

If enough care is not taken the paint can lift away from the chromed panel after a while. When a panel is being re-chromed make sure that only the rim is polished and that the centre is as clean as it can be before painting. When John Marks of Vintage Restorations paints one of these panels, he first masks the chromed rim and then bead blasts the centre to provide a key for the paint. Always use an etching primer first before applying the final colour. Some suppliers, such as Moss, sell cans of paint matched to the original metallic tan colour. The panel was fixed to the dashboard by six small coach bolts that were black painted on all cars. The prototype TA and the first few production models had concealed fixing for the panel and a few cars had only four fixing bolts. Replacement panels are available. The Whitworth Shop, in Ohio, U.S.A., offer an excellent service repainting and screen printing your panel to the original pattern. The TF instrument panel is of a different design but was also finished in the metallic bronze/tan colour. The control knobs on the TF were embossed with letters to identify their purpose so no panel lettering was needed.

MAP LIGHT DF41 Part No. 56040A. 30 M.P.H. WARNING LIGHT DF41 DA21 Part No. 56042A. The bases of these two lamps are identical, only the top cover differs. On the TA, TB and early TC the bases were painted black but this changed to tan when the colour of the centre panel was altered. The 30 m.p.h. lamp was first introduced in the 1930's at the time the general speed limit for urban areas in the United Kingdom was raised to that figure. The idea was that it operated between 20 and 30 m.p.h. to indicate that you were not exceeding the speed limit. This lamp was only fitted to home market cars, its place taken by a second map lamp for export models. Some restorers of home market cars have rigged it to operate as a warning light for flashing direction indicators, although the

The 30 m.p.h. warning light fitted to home market TAs, TBs and TCs.

wattage of the bulb should be reduced to avoid it being too bright at night. The special cover has been reproduced by Vintage Restorations which converts a standard DF41 lamp to a 30 m.p.h. lamp.

OTHER SWITCHES. The PS6 L switch, part no. 314090 was used for the TA, TB, TC and TD panel lights and fog lamp. From TD 10751 L.H.D., TD 10779 R.H.D., the panel light switch was changed for one incorporating a rheostat to dim the lighting if required, part no. 78405. On export TCs fitted with flashing turn indicators a switch, SD84/L18 part no. 31219A, was fitted to the centre of the instrument panel in place of the inspection lamp sockets. The knobs of the starter and mixture controls on the TC are of a different shape to those of the TD. The TC knobs are of a slightly smaller diameter, with a longer stem, than the TD. The TF knobs are different again, being octagonal, and carry letters to indicate their use. TDs, after TD 22315, were fitted with flashing direction indicators for some overseas markets. The instrument panel gained an extra hole in the centre for the WL3/1 warning lamp. Where this was not needed the hole was covered with a plug. This same style of warning lamp was fitted to the EX-U TCs as a headlamp high beam warning lamp in place of the fog lamp switch and on both TCs and TDs as an ignition warning lamp. The TDs fitted with indicators had the self cancelling switch, the same pattern as was used on the TF and MGA, mounted on a special bracket beneath the dashboard. The relay was Lucas part no. 33117 and this was used, in conjunction with a flasher unit, to flash a second filament in the front side lights and to interrupt the rear brake light circuit. On the TD the relay was mounted on the bulkhead in the engine compartment and the body plate moved to the side of the bulkhead to make room for it.

OTHER ELECTRICAL EQUIPMENT.

COIL Q12L. Part no. 45020A. Fitted to TA, TB, TC, TD and early TF. this was painted black; frequently replaced by owners with Lucas "Sports Coils" which were aluminium, often with a browny-red top cover. The later TFs were fitted with LA12, part no. 45172. These had an unpainted aluminium body and black top cover. Coils must be connected correctly with the - SW terminal to ignition switch and the + CB terminal to the distributor contact breaker low tension connection.

VOLTAGE REGULATOR. The pre-war cars were not fitted with a regulator but carried the Lucas CJR3L35 type cut out and fuse box with six fuses. Post-war there were three different types of regulator fitted.

TA/TB CUT OUT AND FUSE BOX

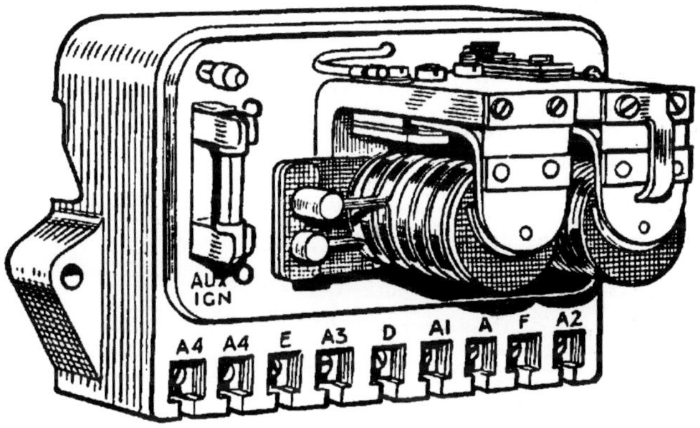

EARLY TC VOLTAGE REGULATOR

The early TC had the large RF91 unit which was replaced by the RF95/2, part no. 37057, from TC 3414. This latter unit was also fitted to the TD until it, in turn, it was superseded by type RB106/1 with a separate fuse box type SF6. All of these units carried only two fuses and for the American market, TC EX-U, an extra fuse box SF5/2, part no. 37107A was fitted.

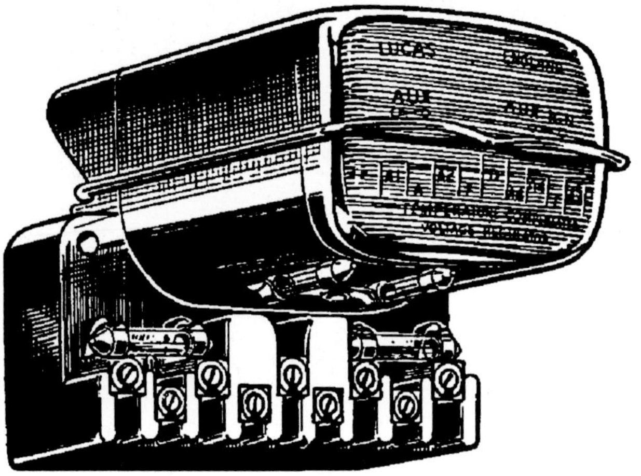

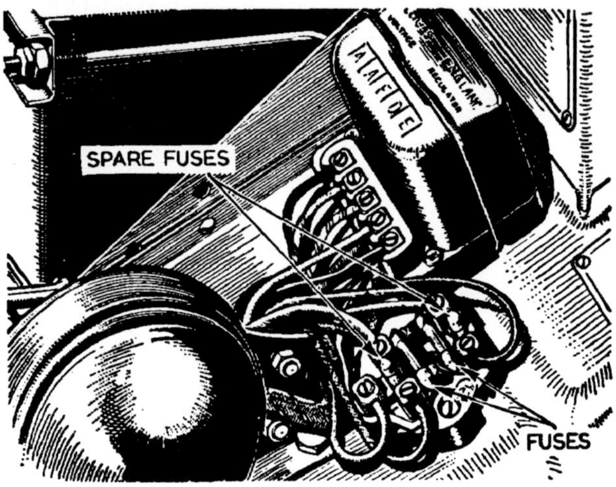

Late TC and early TD regulator and cut out is shown on left and the later TD/TF regulator and separate fuse box on the right.

DYNAMO. The TA and TB had a three brush dynamo, type C45NV2 producing both low and high rates of charge. The TC was fitted with the Lucas C45YV-3 GC24, part no.228334. The early TD had type C39PV DA41 fitted, later replaced by C39PV/2 which continued on the TF. All of these had a drive for the tachometer but there were many similar units fitted to other vehicles that were not so equipped. It is not too difficult to modify one of these to suit a T-type especially if you have the old unit to rob for parts. The commutator should be clean and bright. The surface can be cleaned with carbon tetrachloride and glass paper and the insulation between the segments should be undercut about 0.6 mm. The carbon brushes are easily obtained as are the shaft bearings. Do not over-lubricate after assembly.

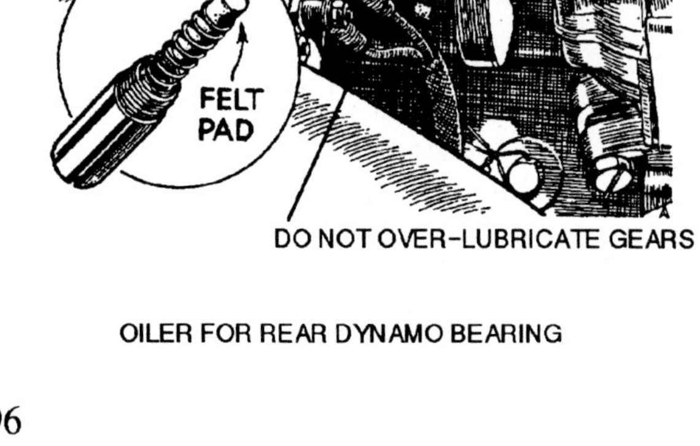

OILER FOR REAR DYNAMO BEARING

STARTER MOTOR. The TA/TB unit was type M418A84. The TC had a Lucas M418G LO starter motor, part no. 255378, and a starter switch ST10, part no.764251. The TD and TF had the M35G/1 L3/1, part no. 25022, fitted. This starter has 10 teeth on the pinion and will not fit a TC unless the ring gear on the flywheel is changed. This later starter is readily available at reasonable prices and it is usually cheaper to obtain a replacement unit rather than overhaul an old one. The commutator can be cleaned with carbon tetrachloride. The later cars had an ST19 starter switch, part no.76701, as fitted to MGAs.

LAMPS. The 8 in. headlamps on the TA, TB and TC for the home market were arranged so that on dip only the nearside lamp remained lit. For export markets both lamps dipped and the later, special EX-U model was fitted with 7 in. S700 units capable of taking sealed beam units. The early lamps, fitted to TA, TB and early TCs, had a "U" shaped pattern in their fairly flat lens glasses with rims that had started with quite a flat section before curving over to hold the glass. The later MBD140 lamp had a much more domed lens with a diamond shape in the glass and rims that have a tighter, more uniform curve. Both types have the inset "King of the Road" badges and are chromed all over, except for the mounting plate. When restoring lamps remove the rivets and take the plate off before having the body re-chromed. The mounting plate was black painted.

TDs all had double dipping S700 headlamps. During the production run the straight dip lamps fitted to all the

Headlamp with "U" pattern lens fitted to TAs, TBs and early TCs.

FT 27 spotlamp. Difficult to find at autojumbles although luckily reproductions are now being made.

The round tail/stop lamp fitted to early TAs. More usually seen chrome plated..

early cars were replaced by units that dipped either to the right or to the left according to the market. There is a lot of doubt as to which TDs had lamps chromed all over. During production there was a period when the body of the lamp was painted, although the rim remained chromed, but details of this are sketchy. Some authorities claim that all Mark lls had chrome headlamps but even this is uncertain. The truth may be that, chromed lamps being in short supply, they were fitted at random as and when they were available. TFs all had the flush fitting F700 lamps.

The front sidelamps were all the chromed Lucas 1130, part no. 52030. They carried a red spot on the top as an indication to the driver when they were lit. This plastic red lens fades badly but luckily replacements are available although these may have to be filed a bit before they will fit. The front glass had the rim formed around it and replacements are produced along with the special stepped screw that holds them in the light body. However, it is possible to have the rim on an old lamp re-chromed without removing it from the glass - speak to a good chroming company.

TAs up to approximately chassis number 1789 had a single round tail/stop/number plate lamp, ST38, mounted on an extension to the number plate. This was chrome plated all over on most cars although on a few early examples only the rim was chromed. To meet modern regulations two lamps would have to be fitted.

Later TAs, TBs and TCs were fitted with a single Lucas ST51 D lamp, part no. 53029, incorporating tail, stop and number plate lights. This "D" lamp had a separating bar in the lens and was chrome plated all over. The EX-U variant was fitted with two high mounted 482/1 (part no. 53131) stop/tail lights held by brackets to the sides of the petrol (gas) tank. On this model the number plate light was a 467/1 lamp, part no. 53093, which was also used for the TD and TF. Early versions of this lamp had two bulbs and the later just one. The rear stop/tail lamps for the TD were Lucas 471, later replaced, for TD 21303 onwards, by the circular 488 type mounted on a chromed plinth. These later lamps were also fitted to the TF.

The spot lamp fitted to TAs, TBs and early TCs was the FT27. This lamp is not now easy to find although some reproduction ones have been made. From TC 4739 the lamp changed to the SFT 462 and this was also available as an extra on the TD. The option available for the TF was the SFT576 often paired with the SLR576 spot lamp. The EX-U TC did not have a fog lamp as standard. Where fitted in export markets the fog lamp was moved to the other side of the badge bar if the car was to be driven on the right hand side of the road.

REFLECTORS were fitted as standard on TF 1500 models and these were mounted on curved rubber plinths to the sides of the body above the rear wing. Reflectors were also fitted to the earlier cars for some overseas markets and are now an M.O.T. test requirement in the United Kingdom.

HORNS. TA/TB horn was a Lucas "Altette" type HF/934/2. The TC was fitted with the Lucas HF1234 horn which was similar to the pre-war item with a chromed outer bezel and a black crackle painted centre. The domed nuts securing the bezel and the centre ring were also chromed - pre-war being fluted and more pointed than the later ones. These horns are very reliable and seldom need drastic attention. Usually failure to work can be traced to poor electrical connections or loose mounting. The Lucas service instructions for this model gives the current consumption as 4 amps., when correctly adjusted. To test and, if necessary, adjust the horn you should connect a D.C. ammeter in series with the horn. Operate the horn and turn the adjusting screw on the back anti-clockwise until the horn just ceases to work. Then turn the adjustment screw clockwise one quarter of a turn and check the performance and current consumption. Further adjustment is then made by advancing the adjusting screw a "notch" at a time until

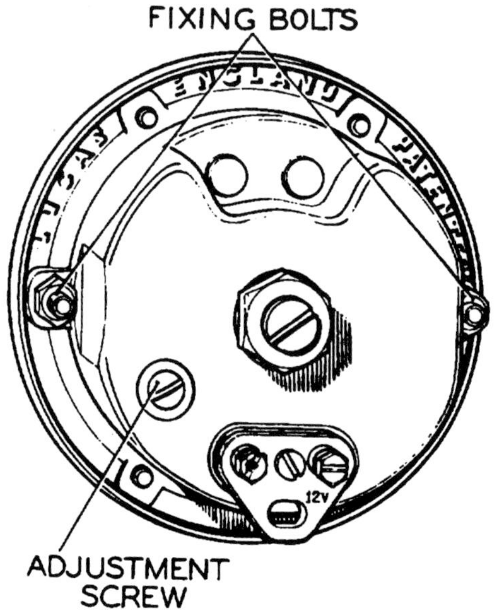

ADJUSTING TC HORN (REAR VIEW)

the best results are obtained with the current consumption not exceeding 4 amps. In the absence of a suitable ammeter turn the screw anticlockwise until the horn just ceases to sound and then turn it clockwise 6 "notches", one quarter of a turn, when the setting should be about right.

The TD and TF had two windtone horns type WT614. The Type WT614 horns were also fitted to the EX-U TCs. The TD horns were mounted on the bulkhead but shortage of space in the TF engine bay forced their re-location to a position below the radiator. To adjust these later horns first remove the cover and then slacken the locknut on the fixed contact and rotate the adjusting nut until the contacts are just parting and the horn fails to sound. Turn the adjusting nut one half a turn in the opposite direction to just close the points and tighten the locknut. If the horn fails to work check and clean the contacts.

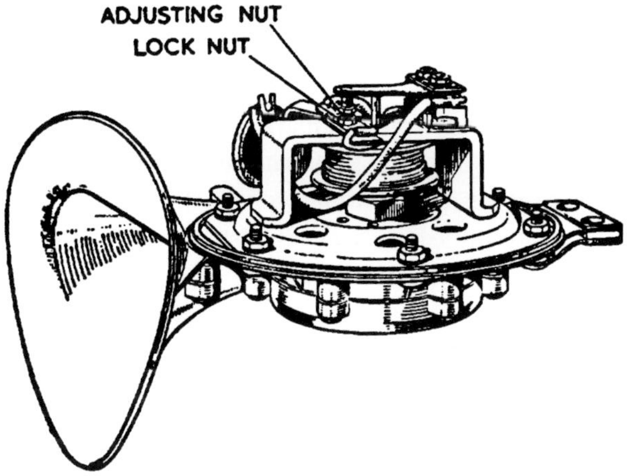

ADJUSTMENTS ON WT614 HORN.

FUEL PUMP. All models up to the early TF have the low pressure SU pump mounted on the bulkhead, the later TFs have a high pressure pump mounted at the rear. The early pumps, including early TCs, had brass bodies and the later aluminium. The TD MkII has twin pumps with the fuel lines duplicated right back to the tank. High engine compartment temperatures, especially on the TF, may have forced the change and today many owners fit rear pumps to the cars to help overcome fuel feed problems when hot. To maintain an original appearance the existing pump can be left in place with the valves removed. If you decide to go this route ask the loom manufacturer to incorporate the pump feed in the loom. Wherever the pump is mounted ensure the earth is good.

Most of the problems with these pumps centre round the contacts. Properly set up and used regularly the pump works well and will deliver enough fuel for even the thirstiest engine. However if it is left unused for long periods the points can corrode and the pump needs a "bash" to get it moving! Cleaning the points will usually help but if more drastic attention is needed most parts are available and the TD/TF manual gives rebuilding instructions. If the points are burned replace the condenser. The fuel filter should be cleaned from time to time, especially if the fuel tank is old. If a separate filter is fitted this should be between the pump and the carburettors.

The TA/TB petrol tap fitted to the side of the toolbox sometimes causes problems because the seals dry out. When assembling the tap soak the cork seals in oil for some days first.

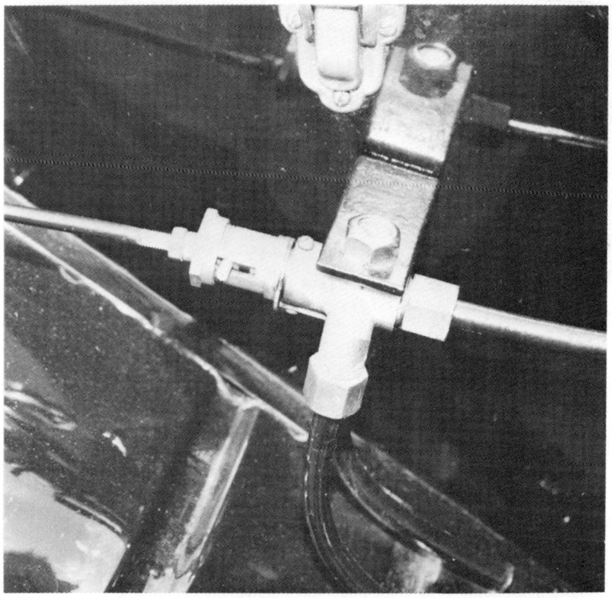

Petrol tap on TA. Twin pipes run from the tank to this tap which is mounted on a bracket bolted to the side of the tool box.

WIPER MOTOR. The Lucas CW1 motor was fitted to the TA, TB, TC and the TD except for the Tickford. This was made for many years in 6, 12 and 24 volt guises. When looking around at autojumbles make sure to buy a 12 volt version. The internal gearing varied from model to model to give different angles of sweep. It is an easy matter to swap these gears to change the size of arc the wipers sweep if buying a motor originally intended for another vehicle. The T-series cars use a 150 degree sweep and this is stamped on the larger gear wheel. Supplies of new motors appear to be drying up at the time of writing this so do not throw anything away. Armatures can be re-wound if burned out and parts from other, similar, motors used to keep yours going. Leaks through the wiper spindle are often due to the felt seal either drying out or not being in place. This seal should be soaked in oil before fitting around the spindle and under the chromed cover on the ouside of the screen. There are also fibre seals under the chromed washers holding the 3 BA fixing nuts. The wiper arms for the early cars had the locking nut securing the arm to the shaft on the top whereas the later cars had this underneath. As listed elsewhere the TDs, from chassis number 22315, had the wiper motor moved to the centre of the screen. I expect this was to avoid the need to transfer the wiper for L.H.D. models but it also removed it out of direct contact with the passenger's head in an accident! The wire for all TDs came through the scuttle on the right hand side and was transferred to the other side of the windscreen along the top rail for R.H.D. models.

The TF used a more modern wiper motor, Lucas CRT12, which was mounted on the scuttle with cable drive to the arms. These arms were brought into use by turning knurled knobs, one in each glove pocket; the driver's side knob also turning on the motor. Spares for the rack mechanism, and the switches, are available.

DISTRIBUTOR. The TA, TB, and TC were fitted with distributors (DK4 AA35 on TA and DKY4A.DA34 on TB/TC) incorporating micrometer adjustment of the timing. This adjustment is part of the clamping bracket but is a fairly inaccurate method of adjusting timing and any changes made using the adjustment should be checked with a timing light. The TD/TF (DKY4A at first and D2 A4 from TD 22735) did not have a micrometer adjustment but the clamp bracket changed with the change in distributor.

On any T-type that has seen much service the distributor will either be worn or have been rebuilt or replaced at some time. The chances are that it will not perform exactly as it should either because of wear or because over the years parts of it have been modified or changed. The settings given in the TD/TF manual for points gap, for example, are different for "early" or "late" distributors and it may not be easy to see which type you have. On the early distributor the lobes of the cam that operates the points are symmetric and looking at them from above they appear to look almost "square" with flat sides and rounded corners. The later high lift type have rounded sides and corners are very pointed. As these cams are interchangeable an early car may have the later type high lift cam. The gap for the early type is 10 - 12 thou. and for the later 14 - 16 thou. If you have a dwell meter the dwell for the early type was 41 -49 degrees and the later 57 - 63 degrees.

The advance curve of the distributor is determined by the springs fitted to the advance weights. With the engine running it is possible to check this with a timing light as long as the position of the 32 degree mark is first calculated and painted on the timing cover (the setting reduced to 24 degrees for later cars). If the distributor is worn and much work on it is needed then perhaps the best thing to do is to leave the job to one of the specialist firms who will be able to both repair it and check the advance curve. New distributors are available and I have used one of these on a TD without anyone spotting the substitution. To disguise the plug-in leads on the cap I fitted the screw type fastenings with the threads turned down so that they just push into the plug type fittings. A dab of glue holds them in place!

CHAPTER EIGHT

PAINTWORK

It is the outside paint finish of a car which is the first thing that you see when you look at it. No matter how well the restoration of the other components has been done, if the paint work is not good, the restoration will be written off by the casual observer as second rate. As well as considering the standard of the paint finish, it is also important to try to achieve the right look for a car that is being restored to original condition. You may not always wish to adhere strictly to the paint colour originally specified for the car when it left the factory and, indeed, you may wish to use a colour never used by M.G. on any of their cars. However, try to use a colour that suits the car and looks "in period". A few years ago the current MGBs were sometimes painted in a rather bright bronze colour which then looked not too bad and similar to the colour a lot of other 1970s sports cars were painted. However, one enthusiast so liked the colour he painted his TD with it - quite a contrast to the red upholstery! The car was well restored but never looked "right" because it was finished in an "out of period" colour. Other T-types painted in non-standard colours are seen regularly and are often admired but usually only if they are in finishes that would have been seen on sports cars at the time they were built.

For myself I prefer to stick as closely as I can to the standard colours offered by the factory as I think the cars look best in these. If you are fitting a trim kit, or leaving the original upholstery and trim, try to choose a colour that would have been specified to go with the upholstery. I have given below the range of colours and trim options listed as standard, but there were some options that never reached the sales brochures, and were probably only sold in small numbers, as well as variations on overseas assembled cars.

COLOUR SCHEMES.

PRE-WAR TA/TB

On introduction the TA was available in the following colour schemes:-
Saxe Blue with blue upholstery
Racing Green with green
Carmine Red with red
Black with blue, green or red.
From October 1936 Cream with red upholstery was added to the list.

In October 1937 the catalogue shows the following schemes:-
Black with biscuit, red, Abingdon blue or green upholstery
Cream with red
Abingdon Blue with blue
Dublin Green with green
Emgee Red with red.

By January 1938 the range was amended with the name of the blue changing from Abingdon to Saxe Blue and Dublin Green reverting to Racing Green. It is probable that only the names were changed, the colours remaining the same.

In September 1938 the introduction of the Tickford model brought some new colour schemes into the range. The standard cars remained as before and the Tickfords were available in the following colours:-
Black with biscuit, red, brown or grey upholstery
Blue with grey
Red with red
Green with biscuit
Duo-Green with biscuit
Maroon with brown
Grey metallic/Grey with grey
Grey with grey
The hoods were available in blue, maroon, fawn, grey and black to tone with the paint colour

In June 1939 the range for the TB was as follows:-
Black with any standard leather colour
Saxe Blue with blue
Coral Red with red
Apple Green with green
Duo-Green with biscuit
Maroon with maroon
Grey metallic /Grey with grey
Tickford colour finishes are:-
Black, Saxe Blue, Coral Red, Apple Green, Duo-Green, Maroon, Light Grey and Metallic Grey/Grey with biscuit, red, brown or grey leathers. The hood was either black, maroon, blue, grey or fawn.

Looking at some old colour charts, which cannot be guaranteed to give exact colour matches, Coral Red was a darker red than Emgee Red and Apple Green was a slightly lighter shade than Dublin Green. However, Saxe Blue and Abingdon Blue look almost the same.

POST-WAR.

Upholstery was available in three colours - red, green and beige. Black cars could have any of these, red cars had red or beige, green cars green or beige, cream cars red or green but blue cars were limited simply to beige. The paint references I have given below refer to the BMC paint codes and are given merely for comparison.

TC. At first it appears that TCs were available only in Black but by 1947 Regency Red and Shires Green were added. In 1948 and 1949 the range was extended as follows:-
Black BK1
Regency Red RD14
Shires Green GN25
Sequoia Cream YL5
Clipper Blue BU14

TD. Initially the range was:-
Black BK1
M.G. Red (same as TC) RD14
Ivory (same as TC) YL5
Clipper Blue BU14
In addition there was Almond Green. There is some doubt as to exactly which colour this was as some references mention that the body was metallic and the wings solid colour. I am sure, however, that although some cars were finished this way (probably with Metallic Green GN31 for the body and Almond Green GN37 for the wings) some were finished in the TC Shires Green GN25.
In 1951 some additional colours were listed but these appeared on only a few cars:-
Autumn Red RD6 (darker than M.G. Red)
Sun Bronze (metallic) also used on Y-Types
In 1952 the range was changed again to:-
Black BK1
M.G. Red RD14
Ivory YL5
Woodland Green GN29 (a darker colour than the earlier greens used, with more blue in the mixture)
Silver Streak Grey (metallic) GR25 - a rare colour.

TF. The colour scheme for the TF was constant throughout the production period and was as follows:-
Black BK1
M.G. Red RD14
Ivory YL5
Birch Grey GR3
M.G. Green. This colour listed in the catalogues appears to have been both a solid colour (possibly Woodland Green GN29 as on TDs) on some cars and a metallic GN31 on others.

As in all things, the owner has the choice of colour when painting his car and should choose something he likes. Most seem to choose either red or ivory if the car parks at M.G. meetings are a guide but as to the *shade* of red or ivory choosen then that is something else! The question of the colour mix used originally is one that will always cause enthusiasts many problems when restoring their cars. There have been attempts over the years to define the original M.G. colour range in terms of paint codes for currently available paints from various manufacturers. The listings of these have appeared in books and magazines in the United Kingdom and abroad but still almost all the cars you see differ in shade from other equally carefully restored examples. Why is this? Well, paint manufacture has changed greatly over the last few years and many of the pigments used even a short while ago are no longer allowed under current safety legislation. This means that any obsolete formula mixed using modern pigments will look different from the original. Also there were quite a number of changes to, for example, M.G. Red (or Emgee Red) during the production run of the cars and a match even then was not always possible. I tried a couple of years ago to use an often quoted I.C.I. code for the ivory colour used on the TC, to find that this code was actually for the Royal Ivory paint used on the Morgan car until quite recently and that this was really far too "brown" a colour for the TC, which had quite a "yellowy" tint to the paint. In the end I opted for a colour mixed from one of their ranges which was a close match for a paint sample on the car.

So where does all this leave you, the restorer, looking for guidance here as to the colour you should paint your nice newly-restored bodywork and wings? The best advice I can give is to either try to match to the original paint, if you can find any, or talk to owners of cars that you see, with a paint colour you like, and try to find out from them the exact paint they used and the manufacturer. I know this does not seem like the best solution but in my experience trying to match old codes is, at best, misleading and you are better employed choosing a colour that looks right to you and staying with it. Whichever colour is selected, make sure to buy enough paint for the whole car. Later mixes may differ from the original batch and it may even be worth having an extra litre or so to keep for future use.

If there is some of the original paint still on the car, and some of the original trim, then there is no trouble finishing the car to approximately the colour it left the works. If there is no trace of the original finish then you will not be able to find out the original colour from any factory records as these have now been destroyed. The only exception to this is the TF. The TF chassis number prefix incorporated a paint colour code in the third letter. These were:- A - black, B - light grey, C - red, E - green, H - primer (for overseas assembled cars) and P - ivory. The British Motor Industry Heritage Trust also have some records from 1955 onwards, which may include the very late TFs, so if you have one of these it may be worth trying them. There are some of the original factory colour charts still around but, whilst they are of interest, they do not prove a very reliable guide to the original colours. It seems that these samples were treated to a sealing coat of varnish when printing and this varnish darkens with age discolouring the paint sample - oh well...!

Having chosen the colour, which type of paint finish are you going to use for your car? Originally most of the T-series cars had the body finished in cellulose and the wings finished in the more durable enamel paint. Nowadays quite a number of cars are finished with modern "two-pack" paint which is very much harder wearing than the original finishes. The only problem with this type of paint is that it is totally unsuited to home application unless you have the necessary paint booth and air fed breathing equipment. Indeed, many of even the "ordinary" cellulose paints now contain quite a cocktail of chemicals and great care should be taken not to breathe them in. Some of the thinners in use are especially potent! As well as breathing in vapour there is also a health risk from skin contact with some of these paints and thinners. All in all I feel the time is fast approaching when legislation will stop us spraying at home and I understand that in some countries there is now a difficulty in obtaining materials for non-trade use. As someone who has had a lot of enjoyment over the years painting my own cars I am sorry that things are going this way.

However, if you decide to paint the car at home you will probably opt for cellulose as this is the easiest to work with. I cannot here give a complete guide to spraying but I will just point out a few things I have learned over the years - the hard way.

EQUIPMENT AND MATERIALS.

There are quite a number of fairly inexpensive compressors and spray guns on the market and elaborate equipment is not needed to achieve a good result. For many years I used an outfit purchased for £40 in 1969 and only changed this last year when I wanted something with a tank so as to have a small reserve supply of compressed air to use in the workshop. A compressor with a tank of at least 25 litres capacity and a moisture separator will be quite sufficient both for spraying and

Original sales brochures issued by the M.G. Car Company are useful sources of details of the paint schemes offered in any particular year. I referred to these when compiling this book as well as consulting other contemporary sources.

It does not pay to get too interested in collecting sales brochures as they can be both hard to find and expensive to acquire!

A simple jig to mount the body on for preparation and spraying is very useful. The one shown here has wheels fitted so it can be easily moved in and out of the garage.

also for other workshop needs although a compressor with a high output is needed to drive air tools or where an air fed mask is to be used in addition to the spray gun. I use a low pressure gun which is very suitable for spraying T-types as these can be reduced to quite small pieces. A mask is also needed, the type with replaceable filter cartridges is best, and rubber gloves to use both to avoid contact with the thinners and also when rubbing down.

Most of the paint manufacturers issue detailed sheets giving instructions and recommendations on using their products. I.C.I. have a very useful guide that shows the thinning ratios, type of thinners to use in various conditions and rules on handling, including safety precautions. Instructions like these are a great help in selecting the right products to buy. Using the correct thinners assists in achieving a good result, especially if when spraying in less than ideal conditions. I find that I need at least three different thinners to cater for both temperature and humidity changes and also for the different materials being sprayed.

There are many grades of "wet and dry" papers available as well as types like "Frecut" which are designed to be used dry without undue clogging. I usually make do with some 120 grade for initial rubbing down of any filler prior to painting, 600 and 800 grade for the primer/filler and flatting top coat, and 1000 or 1200 for final finishing.

PREPARATION.

I know that this is often said but I will repeat it here again:- good preparation is essential if a good finish is to result. You cannot cover up anything with paint as it is not thick enough to conceal even quite small blemishes. The surface must be really smooth before applying any paint. Any areas of pitting or small dents that cannot be removed by panel beating must be filled. The traditional "lead loading" is not all that difficult to do but it does take a bit of practice and the materials are expensive. Good results can be obtained from the various two pack fillers around but watch out for areas that flex, like wings, as some of these fillers can crack. There are some that now have aluminium particles in them instead of chalk but these do tend to be a bit more difficult to rub down and again some tend to crack if flexed.

If using cellulose do not apply any of these fillers over paint. When spraying, the thinners travels through the previous layers of paint and unsightly dips can appear weeks after finishing the job where the thinners has crept under the edge of the filler. It is better to use a cellulose stopper to fill small blemishes that appear when rubbing down at each stage. This may take longer to dry but at least it will be compatible with the paint. However, do not use cellulose stopper too thickly as this may later shrink, spoiling the finish.

With the use of a random orbit sander rubbing down large areas of filler will be a lot easier. Any areas of surface rusting will have to be treated with one of the solutions available but make sure there is no residue left that could affect paint adhesion. Thoroughly clean all the metal and filler with a cleaner such as "Panelwipe" before going on to apply paint. The surface should also be finally wiped with a "tack" rag to remove any dust.

PAINTING.

On both steel and aluminium apply a coat of etching primer before the primer/filler is sprayed. These etching primers are usually supplied in two parts and have to be mixed before spraying. Usually they will not require further thinning but if this is necessary use the grade of thinners recommended. Do not rub down after this.

Apply a good layer of primer/filler. Some of these are too thick to spray well with a low pressure gun if thinned to the manufacturers instructions. Try adding a further 10% of thinners if there is any difficulty. Try not to spray the paint on too "dry". It should appear wet and glossy on the surface right after spraying. These paints use a fast thinners and if using them on a hot day, especially out of doors with sun on the panels being sprayed, they can dry as soon as they touch the surface. This dry, gritty finish is difficult to rub down without removing a lot of paint, and quite a lot of tiny air bubbles can be trapped in the paint causing problems later. Although you do not want the paint so thin it "runs", it should have a chance to flow onto the surface before it starts to dry.

Apply quite a few full coats of primer before even thinking about rubbing down. It is a mistake to rub down too often. Priming coats should not need to be rubbed down more than twice if your surface preparation was good. Often only one rubbing down is needed but do not pass on to the final stages until happy with the result. The secret of getting a good surface for the final coats is to use a guide coat on the primer before rubbing down. I find it useful to have a small quantity of primer of a different colour to hand and will mix some of this into the last coat of the primer to change its colour before spraying. This final coat should just be a light one going all over the panel. When rubbing down any dips or hollows will show up well and it can be seen if any areas are missed. I like to rub the last coat before painting colour using "wet and dry" with a small amount of washing up liquid in the water. Remember to use "Panelwipe" before applying the final colour. Rubbing down is a time consuming and messy business - allow at least an hour for a wing. The primers are not waterproof and will absorb some water and so allow them plenty of time to dry in the sun before applying top coats.

It is a lot more difficult to rub down gloss top coats than primer. For this reason do not be in a hurry to apply paint to any part you are not entirely happy with. The final colour coats will not cover any blemishes and as paint is expensive get the primer right before using any. I said earlier that there will be some difficulty in choosing the right colour. If there is any doubt then ask the supplier to mix a small sample to try. One of the major suppliers in the United Kingdom will let you have a sample tenth of a litre bag of the colour chosen. This is quite enough to spray a reasonably large piece of metal. It is better to try out some paint first before committing yourself to 10 litres of the wrong colour!

The finish achieved with the top coats is dependent on the paint flowing on to the panel without either being so wet it runs, or dry enough to leave a dusty, dull and rough surface. As the paint is first applied it will appear a bit rough but, if there is a good combination of consistency and temperature, the paint will flow to a nice even gloss finish. If spraying outside remember that the sun will warm a panel quite a lot and the top surface of it may be a lot warmer than the more shaded areas. This will cause inconsistent results and perhaps it would be better to spray the final coat under cover if possible. If you do not want to be shut in with the paint fumes try spraying in the garage with the doors open.

As with the primer spray quite a good layer of paint before rubbing down. It is possible to get away with just one rubbing down session before the final coats. Use 800 grade for this, unless the paint is very rough, and use some soap to help wet the surface. After rubbing down the paint should have an even, matt appearance. If there are a lot of pin holes then you have probably been spraying with the paint too thick or with too fast a thinners for the conditions, or even holding the gun too close to the work. As I mentioned, the choice of thinners is important. In high temperatures, or high humidity, use a slow, anti-bloom thinners. I tend to use this also for all final coats as it gives better results.

When satisfied that there is a good surface apply the final coats. These should be applied in one go, making sure that each pass of the gun overlaps the previous one before it dries, so that there is no overspray onto drying surfaces. The paint should be thinned as much as possible without the chance of runs - probably 70 - 75% thinners, unless it is a cold day. If you are good, and lucky, the final coat will only require cutting and polishing. If any paint is needed to be removed use 1000 or 1200 grade "wet and dry" with soap to flatten it before cutting and polishing.

There was a time when all you needed to do with the underside of wings, etc. was to give them some underseal and forget them. Now this is not considered acceptable if you are going to show the car and you will have to consider some finish that is both durable and attractive to look at. If having the car sprayed professionally with two-pack paint, this is quite chip resistant and may be sufficient for most purposes. There are, however, quite a range of chip resistant undercoats that can be sprayed under the final finish and some of these are also suitable for home application under cellulose. No paint finish will resist all attacks by stones and grit so you will have to add the occasional touching in session to your maintenance schedule.

ORIGINALITY - PAINTING OTHER COMPONENTS.

When considering what colour parts of the car should be painted the easiest way to look at it is to consider how it was built. The chassis and running gear was first assembled, and the engine installed, before the body was fitted. Obviously the factory would not want anything attached to the chassis to carry the final body colour as this would then have to differ from car to car.

The bodies, wings, running boards, bonnet panels, splash panels and petrol tanks were all one colour and arrived at the factory ready painted. The door hinges should all be painted body colour. The petrol tank straps, spare wheel carrier, headlamp brackets and, for

The TD production line at Abingdon. Behind the TDs a Y-type saloon is being built.

the TA, TB and TC, badge bar brackets, were also the same colour. The side plates for the petrol tank were body colour, with the rim polished chromium plate. The radiator slats were painted to match the trim colour. The TF slats were all chromed as were a few late model Mark ll TDs. The bolts holding the headlamp brackets to the radiator brackets were painted body colour, although it seems acceptable now to have these chromed. The under wing supporting tube on the earlier cars was also painted body colour - this is often wrongly painted black. No bare unpainted timber was visible. Most was either painted body colour or matt black - including floorboards and the rear compartment panels. The toe board was usually black but on my TC this was definitely originally body colour and I have also seen it painted body colour on some later cars.

The bulkheads on TDs and TFs were attached before the body was painted so should be body colour with the bolt heads painted over. The TC early bulkhead, up to approximately TC 5000, was painted light grey and thereafter was body colour.

All the items fixed to the chassis, axles, shock absorbers, brakes and so on, were painted black. The bolts used were blackened steel.

TA and TB engines were painted red with the bulkhead black. The engines on all post war cars, except early TCs, were painted a sort of dark browny/red. This colour can be bought from most suppliers or, in the U.K., Humbrol Enamel number 20 from hardware stores is a good match. Early TCs, up to around TC 5000, had the engine painted a dark grey with possibly the first few cars with black or dark green engines - opinion on this point varies. The engine colour extended to the block, head, rear cover, side cover plate, water outlet and breather pipe and most of the fixing brackets for the engine mounts, dynamo, etc. The thermostat housing was either red or sometimes black and the bracketry was also sometimes painted black. I have seen cars where almost everything attached to the engine, including even the smaller oil pipes was painted the same colour. On other cars it seems that some of these smaller pieces remained black. The gearbox and top cover were engine colour as was the sump, except for a few TCs where it was left raw aluminium.

The starter motor and dynamo were painted black all over although the reduction gearbox for the tachometer was left unpainted. I prefer to paint these black items in a semi-gloss paint as I think it looks better than a high gloss finish. The air cleaners are black but the pipework is raw aluminium. The pulley on the dynamo was black but the other two were usually engine colour as was the fan although, again, these were sometimes all black!

The rocker cover was painted either aluminium and clear sealed or a sort of greenish-grey colour. There is some debate as to whether the colour actually changed at all and my theory is that they were originally all aluminium colour and that this discoloured with heat to the greenish-grey colour! I do not know if this is the case but I am supported in this view by Al Moss who worked on the cars from new and still owns his original TC (although now long past its third re-build).

The dip stick and gear lever were chromium plated and hose clips, oil filler cap, bonnet catch fittings, gearbox dip stick, TD/TF engine steady bar and late TD/TF turn signal relay were nickle or cadmium plated.

CHAPTER NINE

TRIMMING AND WEATHER EQUIPMENT

Next to the paint finish, it is the interior of the car that most attracts the onlooker. This is particularly true of any open car and with the T-series cars we are fortunate that the interior is particularly attractive. Unlike later cars the trim was never just dull old black! All the cars had leather seats in a range of colours that complimented the exterior finish. As you have probably gathered by now, I favour staying with the original factory options when painting and trimming a car. These colour schemes were, for the most part, well chosen and suit the overall design well. Particularly in hotter climates the original leather would have deteriorated and it is more than likely the car will have been re-trimmed at some time, either completely or in part. Often the seats would have been fitted with some sort of plastic cover and the best of these can look very like leather. I would not like plastic covers on my car but I can see the sense if you live somewhere hot and face the cost of regular replacement of dried up and cracked seats. I know that leather can be treated with hide food to try to preserve it but I have no experience of trying this in high temperatures.

If any of the original trim remains, perhaps buried under a replacement seat cover, this should provide a clue to the original colour. If the original exterior colour is retained then it makes sense to keep the interior original as well. When changing to another colour make sure that the trim shade was available with that car colour. Pre-war, on the TA/TB, there was a larger range of leathers available as can be seen from the lists in a previous chapter. The trim on the Tickford differed from the standard sports model and both brown and grey leathers were available in addition to the red, green, biscuit (tan) and blue of the ordinary cars. You often now see green cars with red trim and ivory cars with tan trim but these were never an option either pre- or post-war. Black trim was also never offered. The hood on the standard car was black until part way through the TC production, when it changed to tan. The hood frames were black painted until the material was changed to tan when the frames were then also finished in the same colour. The Tickford drop heads could be finished in any of five colours, black, maroon, blue, grey or fawn, dependent on the exterior chosen. The material used was very like modern mohair hooding in texture.

The interior panels on the TA/TB were all covered in rexine but post-war the door panels were leather covered until part way through the production of the TC. Probably about the time the dash changed from wood veneer the leather changed to rexine. The top bead of the door panel and the top of the door pockets, however, were still leather. On the TC, and later cars, the skirts of the seats changed to rexine except for the front piece which remained leather. The back of the seat was rexine on all cars.

If faced with a complete re-trim then, unless you have access to an industrial sewing machine, you will probably need to buy a trim kit. There are a number of these available from various suppliers and I have always used those produced by Wm. M. Collingburn of Richmond, Yorkshire. I have found that these usually fit very well and that the colours and style are a good match for the originals. If, however, the body has been completely re-built and there are some doubts about the dimensions, send some patterns so the panels can be made up to fit the car. If the seat bases and back rest are good, just buy seat covers, but if these appear to need any repair then consider purchasing complete seats. This will cut out the chore of fitting the covers to the bases and may even look better than can be managed with only limited experience.

A general note here about using adhesives. Firstly most of them are very inflammable and almost all must be used in well ventilated areas. There are many grades of adhesive available to the trade and a trimmer will use different types for each job. However, one general purpose "contact" adhesive, such as "Thixofix", will suffice except for some panels where a slower drying glue, like P.V.A., may not be a disadvantage. Where P.V.C. or Rexine needs to be glued to a piece of wood or fibreboard, coat the material with a generous layer of glue and then lay the board on to the material. Immediately separate them and set aside for 10 minutes for the glue to dry, before bringing them together again. This will avoid the job of brushing glue on both halves. When gluing edges make sure the glue has dried enough before pulling the fabric over and pressing it in place. Always tack or staple these areas as well. When making corners form a series of folds on the back of the panel and trim the excess fabric with a knife or scissors.

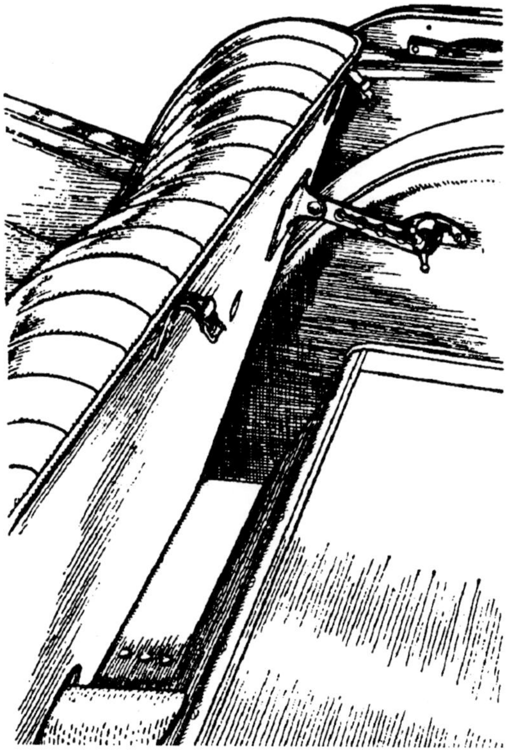

RE-TRIMMING THE INTERIOR - WHEEL-ARCHES AND BACK PANELS.

On all the models the wheel-arch coverings are supplied as two pieces of P.V.C. sewn together to the shape of the wheel-arch. On the TA/TB there is a thin piping at the join, but on the TC and later cars the join is just oversewn, and a slot for the sidescreen bracket incorporated. Kits should also be supplied with felt wadding, to place under the cover, to help to obtain a smooth finish. This wadding should be glued in place before the cover is stretched over and fastened. I usually also glue the cover to the wooden side rails as well as tacking or stapling in place. The inside edge is tacked to the floor of the rear compartment and to the heel board at the back of the cockpit. The bottom is tacked to the tacking strips fitted to the body irons. The edge on the rear floor and back board is covered with a black edging strip tacked in place. The strips attached to the body irons with bifurcated rivets on the TA, TB and TC are fibreboard. The TD/TF have 4 mm ply strips fitted to the metal rear panel in line with the front edge of the rear floor.

Once the wheel-arch covers are in place the rear panels and hood tacking strips are covered. On the TA, TB, TC and TD the hood tacking strips are covered so the fixing screws are concealed but the TF has the strips covered first and screwed to the rear of the bodywork through the covering.

First cut a piece of P.V.C. large enough to cover the centre strip and down to below the top rail of the rear of the body. This strip is fixed under the hood tacking strip, and the strip screwed to the car, covering the edge you have just fixed. Then it is brought over the hood strip and fastened under the rear rail of the body. A thin layer of black felt on the face of the rear body rail acts as padding. The ends of the P.V.C. are turned in over the ends of the hood fixing rail and fastened, and also tacked to the rear body rail where they will be covered by the two side pieces.

The two side curved pieces are not quite so easy to fit. On the TA, TB and TC the covering P.V.C. is sewn to shape but on the less curved TD/TF rails they are cut from the sheet P.V.C. supplied. The trick with the earlier cars is to lightly tack the covering along the seam line to the wood so that the seam is where you want it. Then you fasten the P.V.C. to the underneath of the wood, covering the end next to the rear strip, remove tacks and screw in place on the car. The skirt of P.V.C. is then brought down the inside, after making a neat fold at the inner edge and a tuck at the front edge, and fastened below the line of the rear shelf. Any small bulges and wrinkles you make here will only be seen when the hood (soft top) is up! The fixing bolts for the petrol tank straps go through the rear rail covering and are not covered by the P.V.C. These were originally unplated steel but are usually now chrome plated.

The rear of the TF is finished rather differently from the earlier models. The three rear hood tacking strips are covered completely with P.V.C. before being screwed in place by chromed screws sitting in cup washers. These screws had Phillips heads. There were two P.V.C. covered fibre board panels, one fitted in each of the rear corners and these as well as the covering for the rear rail are fixed in place before the hood tacking strips are screwed to the body.

REAR SHELF AND SIDESCREEN COMPARTMENT.

On the TA to TD the general construction is similar. The rear shelf supporting the hood acts as the top of the sidescreen box and incorporates a stop on the underside for the lid. On the TA, TB and TC there are two shaped side pieces that are P.V.C. covered and fitted using screws and cup washers to the wooden sides of the sidescreen box. The TD has two batons, shaped to incorporate a stop for the sidescreen compartment door. These wooden batons have Tee nuts inserted and after covering with P.V.C. are fastened in place with machine screws from underneath the rear wheel-arches. In all these cars the top shelf is fixed in place with two wood screws and cup washers. The lid of the compartment is 9 mm ply, with plain butt hinges to fix it to the rear floor. These hinges should be secured with Tee nuts and machine screws.

The side panels, or batons, are covered with P.V.C. cut from the roll supplied, and glued using a suitable contact adhesive such as "Thixofix". After the edges are turned over they are tacked or stapled to the wood. The rear shelf is similarly covered with all the edges neatly turned over and fixed underneath. The back edge of the shelf tucks under the rear rail, covering the fixing tacks for the rear rail covering. Two screws secure it to the sides of the sidescreen box on the TA/TC and to the two side batons on the TD. The lid of the box is covered with P.V.C., the edges are turned over and fixed to the back. The inside of the lid is covered with black felt with the pinned edges covered by black P.V.C. strip.

The TF box comprises four metal sections which are covered with P.V.C. and then fastened to the inner wheel arches with machine screws. The plywood lid is also covered with P.V.C. and lined with black felt; it is usual to line the floor of the box with a loose piece of felt. The rear corner panels are P.V.C. covered fibreboard, as is the panel that fits to the back of the body behind the starting handle. These panels are covered to match the rest of the trim.

The lid of the TA/TB box is held in place by two straps, one fixed to each side. On the TC these were replaced by a single strap fixed to the centre of the top edge of the lid fastened by a "Tenax" style clip. The TD reverted to two straps, placed this time on the top of the lid and held by "Durable Dot" fasteners. The TF has a single strap fixed to the front edge of the lid, secured by a "Durable Dot" fastener. The TC style fastener will mark the perspex rear window in the hood when it is left stowed for any length of time and should be covered with some cloth before lowering the hood - unless the car is fitted with a hood having twin rear windows. As an alternative a press stud could be used.

SIDE PANELS AND PIPING.

When purchasing a kit four panels for each side of the car will have been supplied. On the TA, TB and TC three of these are just plain 4 mm plywood panels covered with P.V.C., but the door panel is more complex and difficult to reproduce without an industrial sewing machine. The stitching for the door pocket goes right through the plywood - a difficult job for a domestic machine! As supplied by Collingburn, each of the main panels has "Hidem" binding sewn in place, with the ends turned over the back of the panel. This method seems to have been the one most in use originally, although there may have been some cases where the binding carried round the edges of all the panels as one length.

The door panel should be fitted first after making sure the check strap has been installed. The panel will need to be punched to accommodate the lock screws and handle shaft as well as the holes for the sidescreen fixing plate if the door frame has captive nuts fitted. Check after securing that the door will still open and close well and that the piping round the door is not rubbing on the door frame. The top piping goes from the lock to the rear of the top of the door and is fixed to the panel. The "Hidem" binding is sewn in place with the join under the door lock. The ends of the binding may need to have the cord removed to allow the lock to sit flat on the panel. When the interior panels are fitted, and the piping in place, there should be no painted surface or daylight showing round the edge.

The top edges of all the three side panels must butt up neatly against the piping that runs right round the car. On the TA to TD this piping is supplied in three pieces. The piece with the longer skirt fits from the door lock striker plate on one side of the car, over the top of the dashboard, to the striker plate on the other side. The other two pieces fit from the door lock right back to the front edge of the side hood fixing strip at the back of the rear panel.

Some trial fitting is needed to get the piping to fit snugly once the panels are pinned in place. I use a staple gun to secure the piping to the frame of the car, whilst offering up the panels to try the fit. I find that you should aim to have the piping quite tight against the panel if the stitching is not going to show. I have seen trimmers fix the piping to the panels, rather than the car, as they work along each side, but this seems a more difficult route to me. Fixing the panels with pins, concealed by the "Hidem" binding, is straightforward. Use a flat blade screwdriver to hold the binding apart and a punch on the head of the pin, gripping both in the same hand. This sounds more complicated than it is! Care must be taken not to mark the edge of the binding. When working round the door frame check the fit from time to time - the piping should cover the gap so no daylight can be seen - if you are lucky!

As I said, the piping joins at the door lock striker plate and a small spare piece of P.V.C. will need to be tacked here to cover any gap, or raw edges of piping seen through the cut out in the front panel. Placing all this extra material under the striker plate will alter the fit of the door and this is a point to bear in mind when fitting the lock as the body is built. Perhaps a couple of pieces of thin card should be placed behind the plate when originally installing the lock - these can be removed when the car is trimmed. As an alternative the door lock can be spaced away from the door. Some of the suppliers, Abingdon Spares in the U.S.A. for one, have ready drilled plates for this purpose. They have the advantage that you can easily remove them at a later stage if the door loosens up.

The TF piping is the same as far as the dashboard. There is a crash roll at the top of the scuttle and the piping along the edge of this is thinner than the rest of the piping on the car. There are two short lengths of the thicker piping that run from the door lock to the start of the crash roll. On both the TD and TF the two panels each side of the foot wells are made of fibreboard so it can bend round the scuttle support hoop. The first few TDs lacked this hoop and the panels for these were the same as those on the TC.

THE DASHBOARD.

On later TCs and all TDs the dashboard was covered with Rexine or P.V.C. to match the rest of the interior. The dashboard is 12 mm ply, exterior grade, and the TD has a cubby hole lid cut from the same material. Given a jig or band saw it is easy to make the panel yourself, although I take the easy way out and buy one ready cut. All the fixing holes should be cut out before covering and the board fixed in place to check clearance for the edging and piping. Check that the instruments fit the holes if the panel was purchased ready cut - they didn't on the last one I bought and this was only discovered later.

If the original chrome edging for the dashboard is used this may need to be re-plated and polished. When buying new make sure that it is the correct pattern. The edging comes with different widths of beading and the thicker type does not look very good on TDs. The excess of chrome round the glove box lid does nothing for its appearance. I found a good set, already drilled and shaped, from Abingdon Spares, although I expect other suppliers can also help. I have found that the "contact" type of adhesive is unsuitable for the dashboard. When parked in the sun the high temperatures reached can cause the covering to come unstuck. The best solution is to use a P.V.A. woodworking adhesive, like Evo-stick Resin W, which will work well provided the P.V.C.being used has a woven base to which the glue can adhere. Some of those with a "plastic" surface on both sides would not stick with this glue. Using the P.V.A., apply the glue liberally to both surfaces and, after smoothing out any bubbles or wrinkles, place the dashboard under a weight for 24 hours. Take care that there is a good smooth surface against the face of the dashboard as any imperfections will transfer themselves into the fabric. When it is dry the surplus can be cut away. If this gluing is done well there will be no need to allow an overlap at the edges under the chrome trim. This will be an advantage if the P.V.C. is thick and an edging with only a small overlap is used. However, if a contact adhesive is used, or there is any doubt about the strength of the gluing, overlap the edges so no unsightly edges can show later.

The dashboards are secured to the rear dash panel with wood screws. The TA, TB and TCs with wooden dashboards used slotted number 8, chromed wood screws 1.25 in. long with raised countersunk heads. The later TCs and early TDs used chromed slotted number 10 screws 1.25 in. long with a raised countersunk head and a plated cup washer. The later TDs used the same size screws but with a Phillips cross slot in the head. Unfortunately these are not so easily obtained as most modern screws of this type have "posidrive" heads and I would rather use the slotted type if the old Phillips head is not available.

The crash roll on the TF is made up of three pieces. There is a wooden backing, a rubber padding strip and the cover material. Originally this was P.V.C. but some trimmers use leather for re-covering as this has more stretch than P.V.C. and can be worked more easily.

Wheel-arch cover is glued and pinned, or stapled, to frame. The inside edge is neatened by tacking a black edging strip in place.

On all models except the TF the rear hood tacking rail is covered after it is fixed to the rear of the body. A strip of P.V.C. is first fastened to the underneath and back face of the rail with glue and tacks or staples.

However, Mike Collingburn says that he prefers to use P.V.C. as the thickness of the leather can cause problems where there are a number of layers overlapping one another at the back. If leather is used the wood could be shaved away a bit at the back to give more room and allow the crash roll to bolt down tightly.

To do the job, lightly glue the padded roll to the wood first making sure the fixing bolts are in place. Coat the back of the material with glue and also the back of the wood strip. When the glue has dried lay the material face down and put the wood and pad in place on it. Work the edges over to the back eliminating wrinkles and creases as you go. When satisfied with the finish, trim off any excess and tack, or staple, in place. When working with leather try not to pull it too hard in any one place as it can crack or tear, especially if it has been around for a while and has had time to dry out.

SEATS AND SEAT COVERS.

If new seats were purchased complete then they will only need to be fitted with the various brackets and runners before installing them in the car. If, however, only seat covers were bought now the troubles start!

When removing the old covers do not discard them straight away as they will yield many clues on how to fit the new ones. On the TA to TD the seat back is of largely similar construction. There is a plywood back board with a strip of wood attached to the top. This must be in good condition and not damaged where the

The rear hood rail is screwed to the rear of the body, after folding the ends round and trapping them under the rail.

The P.V.C. is pulled over and pinned under the rear rail of the body. The surplus material is cut away.

tacks go, or splitting apart. If it is stop now and make or buy a new board. The spring casing should also be in good condition, without any broken springs or any stray pieces of wire that could damage the covers. New cases are available if the old one is bad. The spring case is held to the board with scraps of leather tacked in place. Over the spring case there is a layer of rubberised hair which I cover with a piece of sacking or coarse calico. Some of this wadding may have disappeared over the years but fresh supplies can be purchased from trimmers.

Once the shape of the seat back is right the cover can be fixed in place. The pleats are made so that they will sit better when flattened down laying in the correct way. You will see that the leather is folded over to form the pleat and that it will lay down better one way than the other. With the cover pulled over the seat and roughly in the right place count the pleats and make sure that the centre one is in the middle of the top rail and lightly pin in place. Then loosely pin the lower corners and, keeping the pleats straight, next pin the top corners. Pull the sides back and, making sure there is enough material to turn over the top, start to pin the leather in place along the top rail keeping the pleats even and laying down the right way. Looking from the back of the seat this usually means working from left to right. Only lightly pin at this stage in case adjustments are needed. With the top fixed in place work the cover over the seat back evenly and tack the other edges down. Resist the temptation to pull the cover too tight without first checking that it will still touch the seat bases. It is

On TA/TC bodies the curved corner piece has its covering sewn to shape first. This line of stitching has to follow the edge of the strip and I find it easier to pin this lightly in place before fixing the covering underneath. When the glue is dry the pins are removed to give access to the top of the strip.

The strip is screwed in place, trapping the ends under the wood. The corners are folded and the skirt of fabric is pulled down and pinned below the level of the rear shelf and where the tacks are covered by the side panel. The wheel arch cover should be in place first!

easy to pull the back so tight that a gap appears between the back of the seat and the rear of the seat bases. When the cover is attached to the board all round, the bottom projecting plywood on the later TCs and TDs is covered with pieces of P.V.C. This is fixed concealing the top line of pins as described below for the seat back. The back of the seat is covered with a piece of P.V.C. On TA/TB seats this was often black, rather than the same colour as the rest of the interior. To conceal the pins the top of this material is pinned from the back, through a fibreboard tacking strip, and then folded down to cover the back of the seat. The rest of the edges are turned

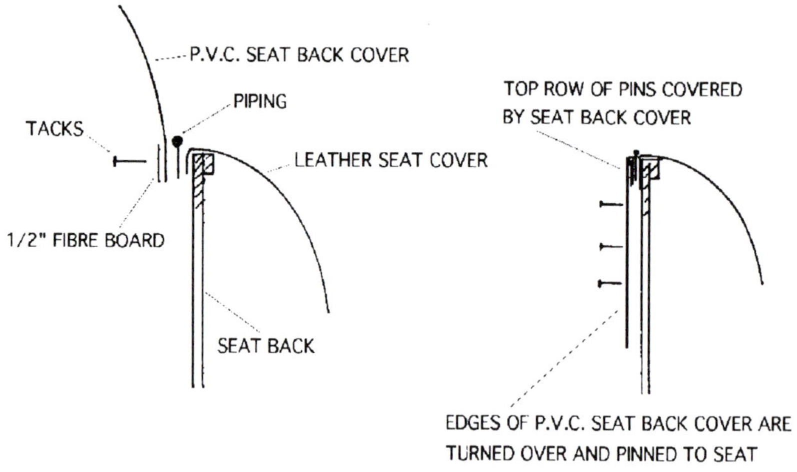

Method of fixing P.V.C. to seat back

Rear of a TC. The lid of the sidescreen box and the two small panels either side are covered with P.V.C. The side panels are screwed in place with wood screws in cup washers. The rear shelf fixes with two wood screws to the top of the sides of the sidescreen box. The inside of the box is lined with thin black felt as is the inside of the lid. The edges of the felt on the lid are covered with a black strip, pinned in place.

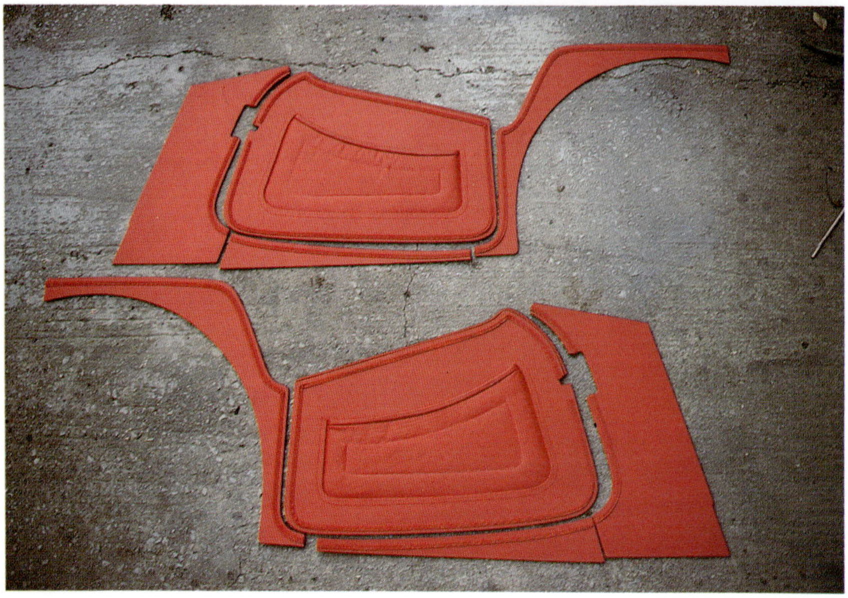

A set of side panels. The "Hidem" binding is sewn in place on the panels.

under before tacking with coloured pins. Along the top edge a piece of piping is trapped between the leather and the P.V.C. panel.

With the seat trimmed, the tonneau cover rail and the seat adjusters are fitted. Tee nuts are fixed to the back board for the adjusters on original seats, and on those I have been supplied by Mike Collingburn. The rail should be painted black, as should the pockets for the seat hinges.

The seat bases are usually in poor condition as damp has often caused the plywood to delaminate. New bases and foam rubbers are available but choose carefully as some you see are far too soft and will be very uncomfortable on a long run! The plywood base should have a wedge fixed to the front to lift the foam up to give support under the knees. Fixing covers to the seats is not difficult. Remember to make sure they are fitted on the correct sides - the join between the leather and the P.V.C. is on the edge next to the gearbox. After tacking in place the edge of the leather under the base is covered with flat black banding, pinned with tacks. The seat bases should have Tee nuts inserted and the runners are fixed to these with machine screws.

The construction of the TF seat is quite different to the earlier models. The basis is a metal frame to which is attached a wooden, wedge shaped box to give support under the knees and wooden strips to which the seat covers are tacked. The back of the seat has a spring which is attached to the frame at several points. If you

The piping is fixed to the edges of the body, checking at frequent intervals to make sure that it fits snugly against the trim panels.

A screwdriver blade is used to open up the "hidem" binding, and the pins are driven home with a punch and hammer.

still have the original covers look carefully to see how they were fixed before dismantling. The seats should be checked first to see if any repairs are needed. Often some welding is necessary and the wood may need re-fixing to the frame. The seat base has a dense foam squab. The seat back is padded with rubberised hair, the edges being covered with felt. The cover should be trial fitted to see that sufficient wadding is present to produce the right shape. It is important not to over-pad as you need to be able to sit into the seat and not slide around on top.

The seat cover is pinned to wooden tacking strips riveted to the back of the frame and the underside. First lightly pin the centre pleat to the centre of the top of the seat back then work outwards, pinning to the wood strips, checking all the time that the cover is straight Once this is all in place the rear fibreboard panel is tacked by the pins sewn into the piping fixed to the edge of the panel. At the lower edge strips of P.V.C. from the rear panel are drawn down under the seat and pinned.

Transfer the position of the pleats on the seat back to the foam on the base and draw the cover for this over the foam and tack to the underneath, keeping the pleats in line with the marks made on the foam.

CARPETS.

The carpet on all the standard T-series cars was black Karvel. This was quite thin with the pile bonded to Hessian which was a cheap way of making carpet.

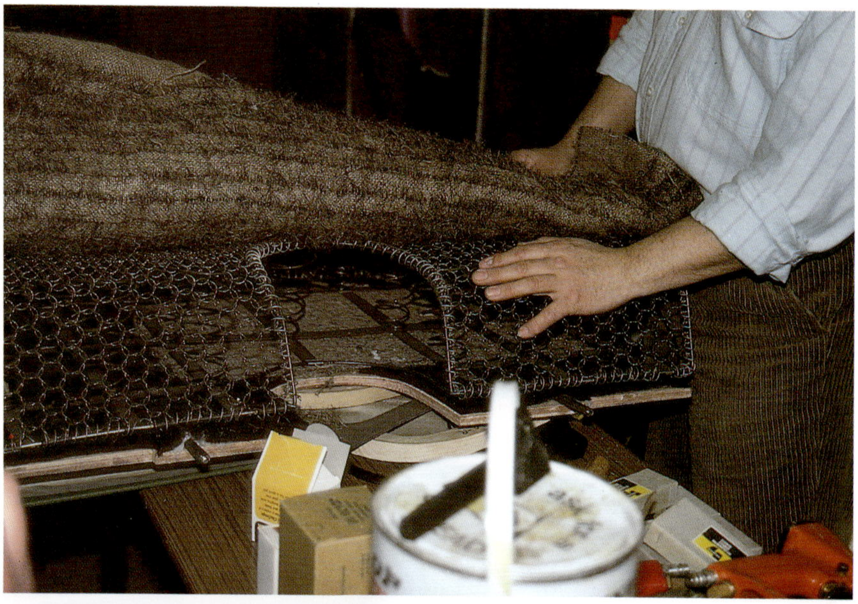

Rubberised horse hair is placed over the spring case and covered with calico.

Mike Collingburn giving a demonstration of how to fix a seat cover at one of the excellent M.G.C.C. T Register rebuild seminars.

As a break from all those detail shots a few pictures of complete cars! The first is a 1937 TA with narrow rear wings. This was my first T-type and my first re-build - nearly 24 years ago!

The second picture is also of a TA but this time something rather different. The special body was built by Park Ward in 1950 as a styling exercise.

This is a 1949 TC recently re-built. The colour, ivory, looks good on T-types but funnily enough white does not. The radiator slats are correctly painted the same colour as the upholstery.

The Bertone bodied Arnolt TDs in open and closed form. These attractive cars were bodied in Italy for sale in the United States. The two cars in this picture by Richard Monk are owned by John Shute.

A later TC with fabric covered dashboard. The Brooklands steering wheel was not standard on a TC and neither was the centre switch for the direction indicators. Where these were fitted on the EX-U model the switch was placed on the instrument panel instead of the inspection lamp sockets.

The more domed, headlights and SFT462 spot lamp on a later TC

TF SEAT

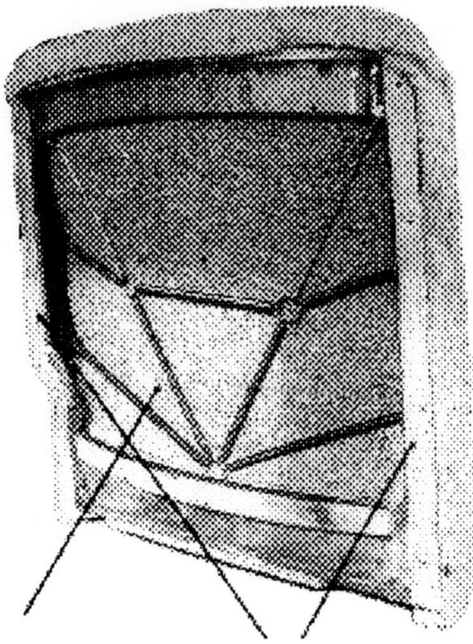

Some of the modern replacements are a good match so long as the edges are not bound. Originally only the edge of the handbrake cable cover on the TD/TF was bound and this piece also had a stiffener of fibreboard. My feeling is that binding the carpets "dresses up" the interior too much - but that is only an opinion. It is easy to cut out the carpets from the roll but as sets are not too expensive I buy mine ready cut as it saves time and also saves trying to work out how much carpet to buy. The sets I have bought from Mike Collingburn have all the correct fasteners supplied, although I do not always fit them exactly the way they were originally! There was no carpet fitted behind the seats but now most people carpet this area. On the TF there was a strip of carpet fitted across the heel board behind the seats, the end of this was bound to match the trim colour where it covered the wheel arch. Early TAs had a carpet over the gearbox and this was held in place with "lift the dot" fasteners, later cars, along with the TB and TC, had a rubber cover here. The leather gaiter around the handbrake on the TA, TB and TC was black as was that around the TD/TF gear lever.

Under the dashboard on the TA to TD there was a panel made of plywood or fibreboard, covered in rexine or P.V.C. Kits containing patterns are available and they are easy to make and install. Do not fit them, however, until you are sure everything works well. The TF under dash panels are shaped fibreboard.

FITTING THE HOOD (SOFT TOP).

Ignoring the Tickford, which has a much more complicated top and rather outside the scope of this book, it is not difficult to fit a hood to a T-series car. The best fitting hoods are undoubtably those made by an expert trimmer to fit that particular car. That said, a hood purchased from a good supplier can be sucessfully fitted by the home restorer.

The first job is to fit the hood frame, after this has been restored and painted. It may be difficult to find out exactly where the frame should fit on a new body and a certain amount of trial and error will be involved.

A TF seat from underneath.

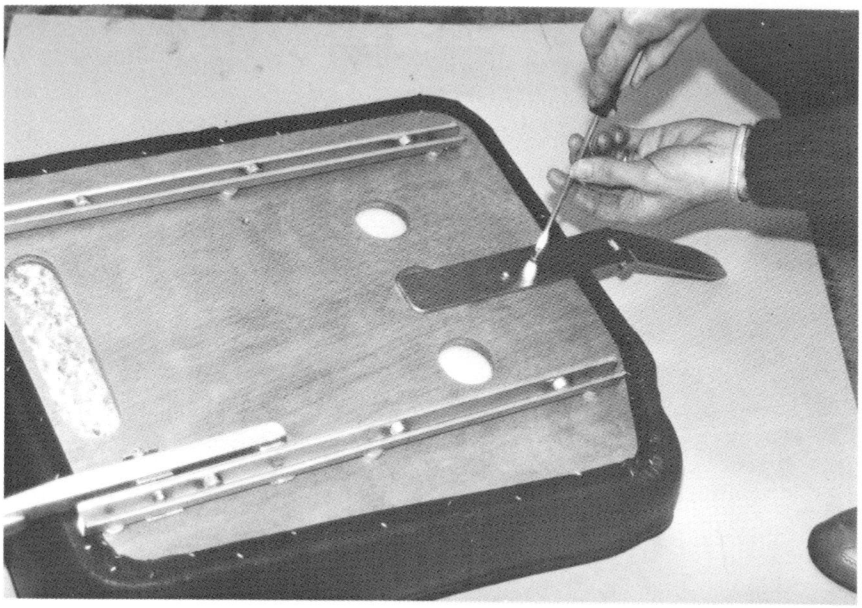

The seat runners and hinge are fixed with machine screws to tee nuts in the seat base.

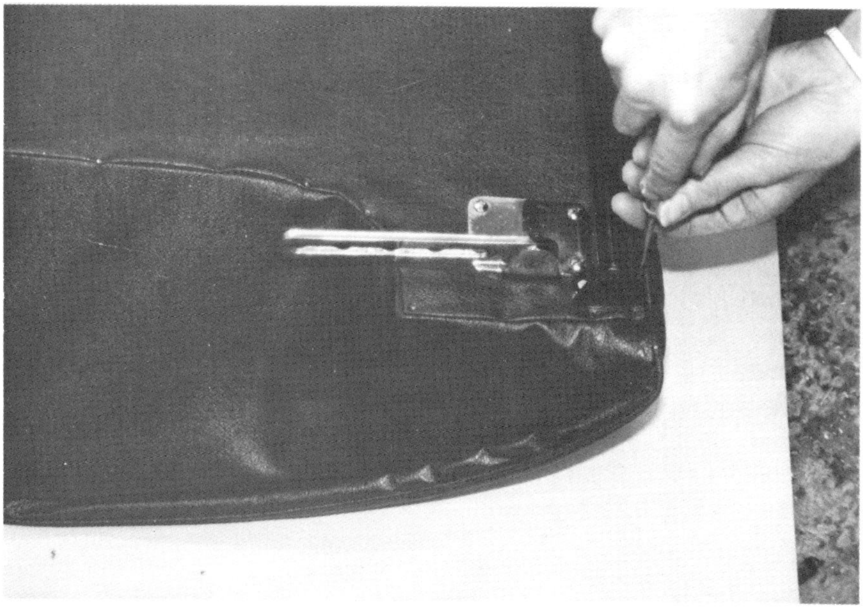

The tonneau cover rail on a TC. The correct colour for the rail is black.

The tool box is lined with white felt, stuck with a contact adhesive. The felt covers the wooden block in the bottom of the box which holds the jack in place. This is a TC.

First fix the wooden front rail to the frame, after checking to see it has the correct curvature and that it fits snugly on the top of the windscreen, then clamp the frame to the screen pegs to establish the correct front position of the frame. The best fit between hood and windscreen is obtained when the wooden rail curves slightly less than the top of the windscreen. It should touch at the centre but sit just above the frame at the ends so that clamping the ends down with the screws holds it firmly to the windscreen all the way across.

Position the side plates to give the maximum headroom at the main bow and fix the plate with one screw each side. Look to see that the frame is level and the same height on each side of the car and that it will lay down inside the body when folded. If correct fix the second screw each side. It is better to use stainless steel screws here, rather than chromed brass, as these are not really strong enough - also make sure they are not too long or the outside panelwork on the body may be punctured.

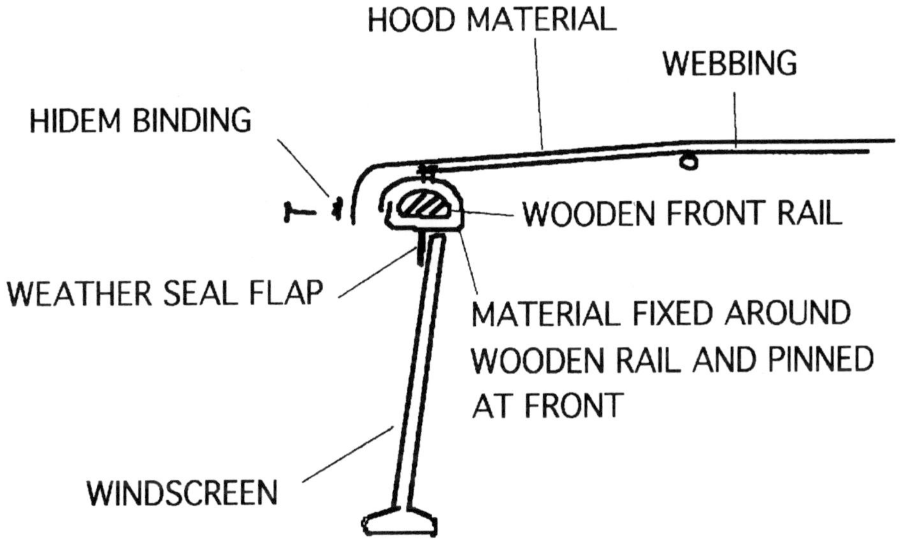

As supplied the hood consists of two parts; the main hood and the covering for the front rail. The front rail is tackled first. This is covered completely with fabric and the piece supplied is sewn to form a weathersealing flap that is notched to accommodate the wiper arms. On some hoods these cut outs are already in place - others you have to cut them yourself. The fabric can either be just pinned in place, or glued to the wood as well using a contact adhesive - I favour the latter.

Coat the wood and the back of the fabric with glue and allow to dry. With the frame erected but not screwed to the windscreen, position the fabric so the cut-outs for the wiper arms are correct and the flap will lie along the front of the windscreen frame when in place. Cut the holes for the pegs, lower the hood frame down and tighten the butterfly screws. When sure the fabric is correctly positioned, work it round the frame after cutting slots to go round the two side frames. Bring the fabric round to the front of the rail and pin in place, trimming off the excess and folding in the ends. The two holes for the pegs are neatened by adding the copper finishing sleeves to prevent wear from the windscreen pegs.

Two inch wide webbing, black for black hoods and tan for the rest, should be fitted to locate the hood rails in place and to help brace the hood material. Two strips of webbing are fixed from the top of the front rail to the rear hood tacking strip on the body. The strips should line up with the bolt holes in the hood frame bows and be fastened front and rear with tacks, and to the hood frame with small nuts and bolts, with cup washers under their heads. The rear hood bow also fastens to the hood fabric on the line of the stitching above the strip over the rear window. The rear hood bow should also line up with the corner of the rear sidescreen frame.

Establish the centre of the front rail and mark this with chalk, doing the same for the rear hood tacking strip on the body. Mark also the centre of the hood, both on the inside and out. Lay the hood in place and fix the centre of both the front and rear with drawing pins, or similar, fastened to the wood strips. If it looks to fit properly fasten all along the back rail with tacks or screws in cup washers, keeping the hood straight and lined up with the bodywork. In some very hot climates the rear of the hood is sometimes fitted with "Durable Dot" fasteners - it can then be lifted up for added ventilation when running with the hood up to protect the top of the head from direct heat. There are two fasteners that fix to the curved side pieces.

With the rear fixed, pull the material as tight as possible and tack to the front rail. Work from the centre, progressing both sides together, until reaching the ends. The sides of the hood need to be pulled round on to the front rail enough to tighten them up and to remove any "waves" in the line down the side of the hood. Cut and fold these ends to make it all sit down well. Once happy with the result, trim off surplus material and cover the tacks and the raw edges with a strip of "hidem" binding. Use a flat bladed screwdriver to open up the binding and a punch on the heads of the tacks. The ends of the binding are neatened up with metal caps. Applying some sort of sealer to the join, before covering it with the binding, will stop water creeping under the edges of the hooding.

SIDE CURTAINS.

These should not be fitted until the hood is in place and the top edges can be lined up with the hood. As supplied the curtains come with the windows covered with material. This is only cut away once the curtains are fitted to the frames and the line of the edge of the frame can be established.

Once the frames have been cleaned and painted they should be fitted to the car and twisted, if necessary, to align with the hood. The curtains are then offered up and the position of the frame chalked on the material. The front curtain should slightly overlap the rear to keep the interior as watertight as possible. Make sure the front curtain will fit well against the windscreen frame and not overlap the side. The tops tuck under the outer flap on the side of the hood and up against the inner flap. When satisfied with the fit, remove the frames and transfer the position of the holes in them to the curtains. Punch holes through the curtains and assemble them to the frames with the special fittings

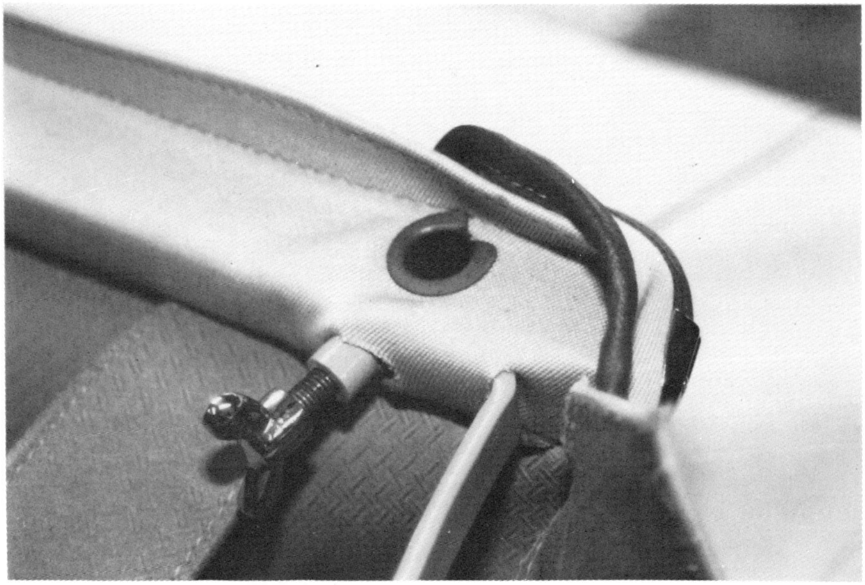

A copper finishing sleeve fitted to protect the hooding from the windscreen peg.

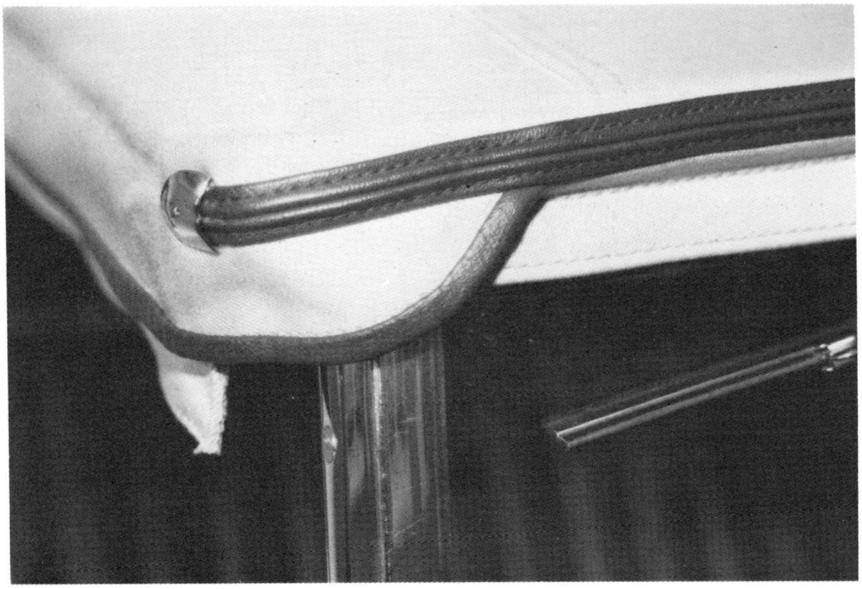

The front corner of the hood is pulled tight and fastened under the "Hidem" binding. Originally the piping and binding were the same colour as the hood material.

The twin rear window fitted to TAs, TBs and early TCs. Although looking attractive they are rather small if travelling much with the top up.

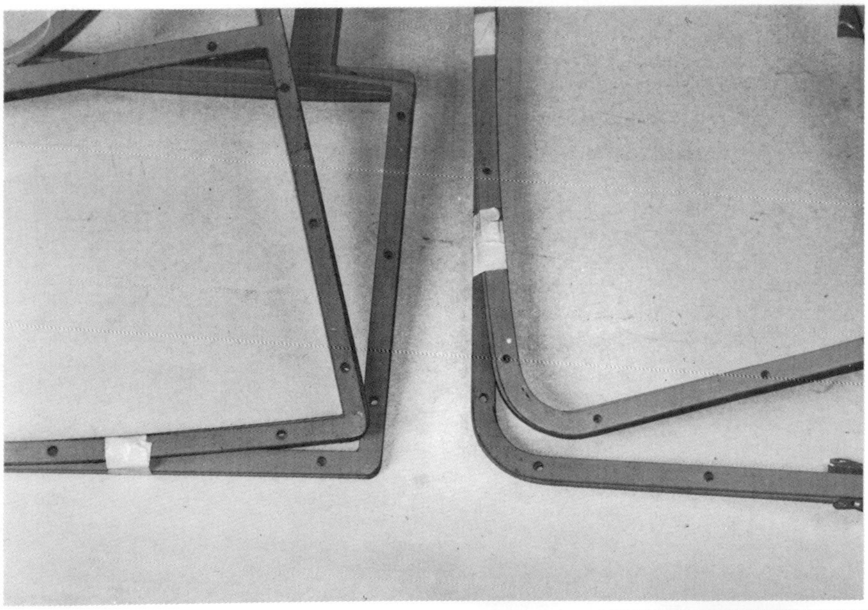

The difference between the sidescreen frames for TDs with two bow hood frames (on left) and three bow (on right). Taller drivers appreciate the greater headroom of the later three bow soft top.

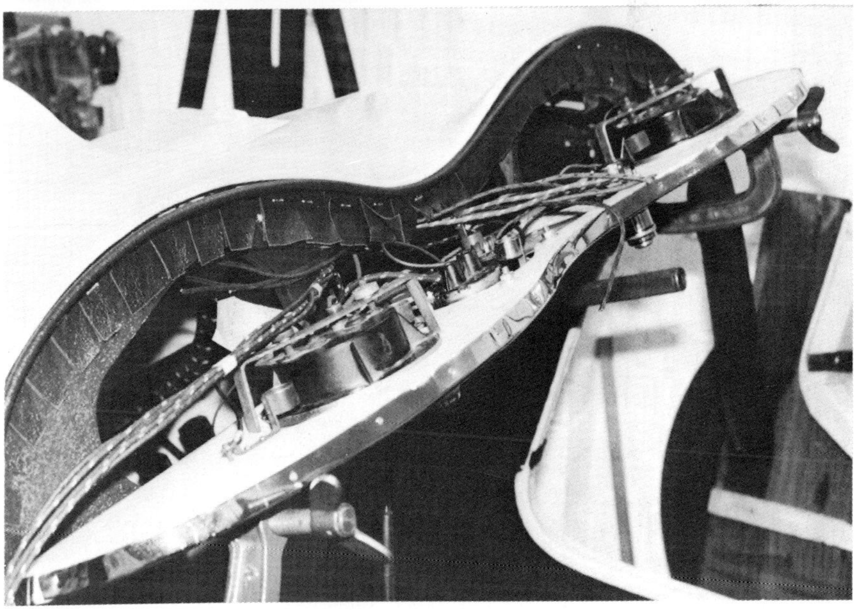

The dashboard can be supported by a couple of "G" cramps clamped to the inner dash frame whilst the wiring is connected. The piping around the top of the dashboard is tacked or stapled to the inner panel.

that also retain the outer chrome strips. The chrome strips on the TD/TF have rounded ends and do not touch each other. Those on the earlier cars are neatly mitred and originally there were small shaped aluminium plates fitted inside each corner to keep the strips aligned. With some of the modern hood materials there may be difficulty in getting bolts long enough to go through all the layers of fabric plus the frame - check on this when buying them. The inner fastenings were just plain thin nuts originally and not domed nuts as are sometimes seen. When the curtains are fitted to the frames, cut away the cover material with a sharp knife to reveal the window, taking care not to cut or scratch the perspex.

There were two small rubber pads fitted to the hood frame to stop the sidescreen frames rattling against it.

TONNEAU COVERS.

Originally all the cars were supplied with a cover for the rear compartment but no full cockpit cover. This rear compartment cover was black rexine on the TA, TB and on the early TC. Once the hood on the TC was changed to tan duck the tonneau cover was made of the same material. The TD and TF had tan duck covers. It is usual now to make the full tonneau in the same material as the hood and sidescreens, usually either double duck or mohair, but there may be a case for making it out of "Everflex" hooding if the car is to see much use in wet weather. Mohair and duck tonneau covers mark easily and are not as weathertight as "Everflex". As the full tonneau covers were not supplied as standard equipment, many owners had covers made from rexine fitted.

The dashboard should fit snugly against the piping, trapping it to the edge of the scuttle. This piece of piping finishes at the door striker plate. As supplied in trim kits the piping has a wider skirt than the other two strips provided to fit down the sides of the car.

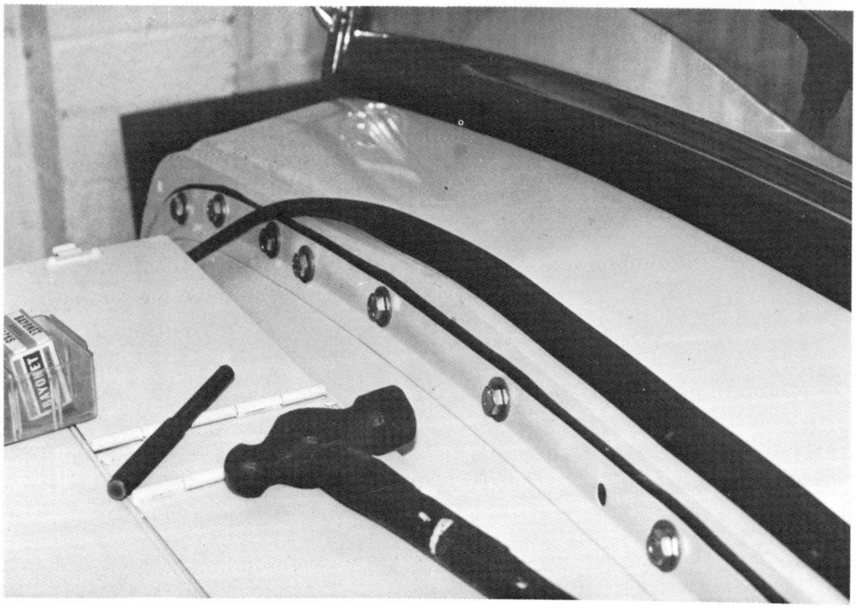

The bonnet (hood) rest rubber is not tacked in place until the fit of the bonnet is checked, as it is possible to adjust it to cover small gaps in the fit between the bonnet and the scuttle.

CHAPTER TEN

ORIGINALITY

I will concentrate here on a fairly disjointed selection of areas where I know difficulties can arise when restoring a car which has lost a lot of its original identity, following changes made by a succession of owners.

The one area where a lot of restorations run into problems is in the engine compartment. Over the years there have often been many extra holes drilled in the bulkheads of most cars and it is necessary to establish early on which of all the many holes were added by later owners. I show later drawings of typical layouts for each of the models. Most of the research for these was done by members of the New England T Register and published in their excellent publication "The Sacred Octagon". There may, however, be some changes over the production period that have not been well catalogued and so do not take these drawings as gospel. The main secret of restoring a car is to make it look right. The exact location of every wire is important but not everything. It is better to finish the car and enjoy it rather than spend all your time worrying over that odd hole!

As the chassis is not immediately visible it is sometimes difficult to see exactly where all the clips etc. fit on a car in the process of being restored. Contemporary photographs and pictures of restored cars are of no help. I have tried to show here some of the pipe runs and positions of clips. Where "P" clips are mentioned these are usually of pressed steel and have swaged edges. These clips are now once again available in a variety of sizes and are usually described in terms of the diameter of the pipe they hold and the size of the bolt hole.

When new these cars were supplied with fairly comprehensive tool kits and on the TA and TB, except for very early cars, these were fitted in rubber trays placed in the top of the tool box. These trays were for many years unobtainable but recently they have again been reproduced. Searches of autojumbles will yield period tools to fit in these boxes and on the later cars these can be placed in tool rolls. Mike Collingburn has produced some tool rolls made to the original pattern.

SALES LITERATURE.

The M.G. Car Company, as well as overseas importers, produced a number of attractive brochures to help sell these cars. It is interesting to try to collect a few of these items to display, either in the garage with the car or at meetings. Some are rarer than others but most were produced in reasonable numbers and are fairly easily located through specialist dealers in such material. A word of warning - some have been copied and these look very close to the originals.

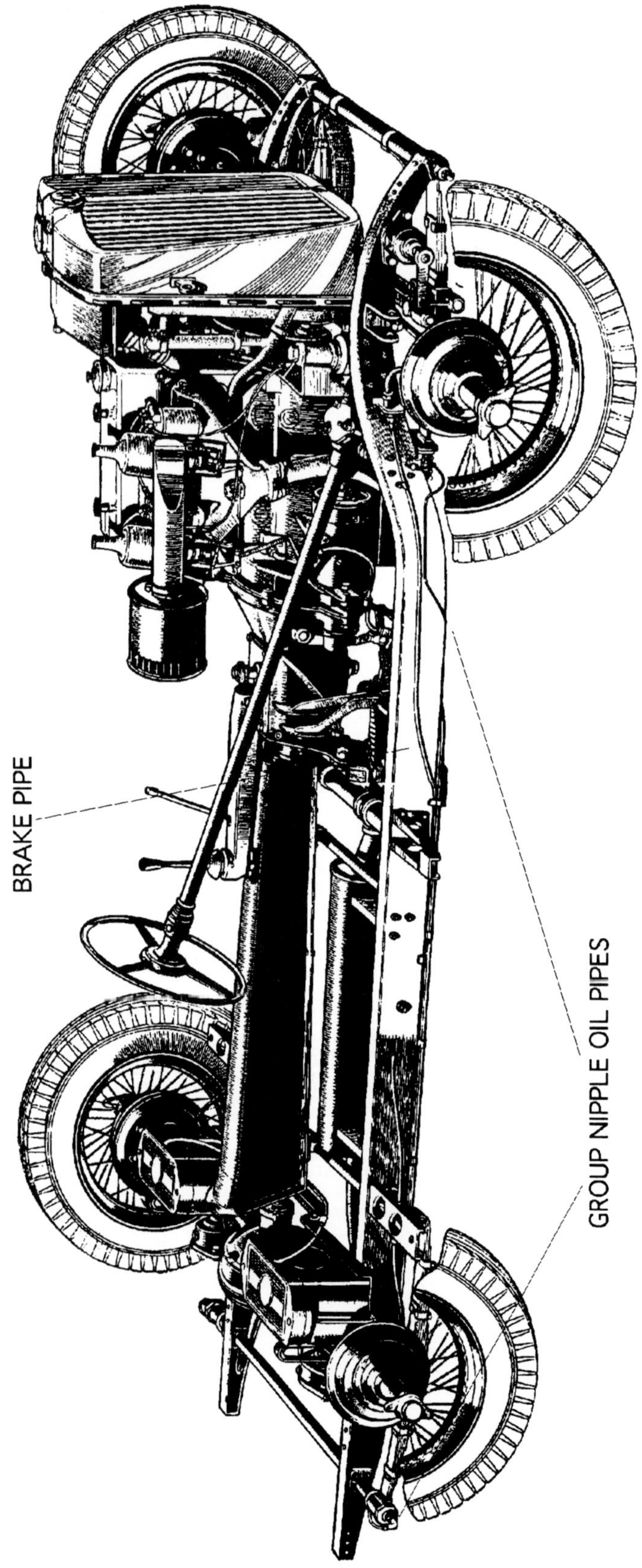

TA CHASSIS SHOWING POSITION OF SOME PIPE RUNS.

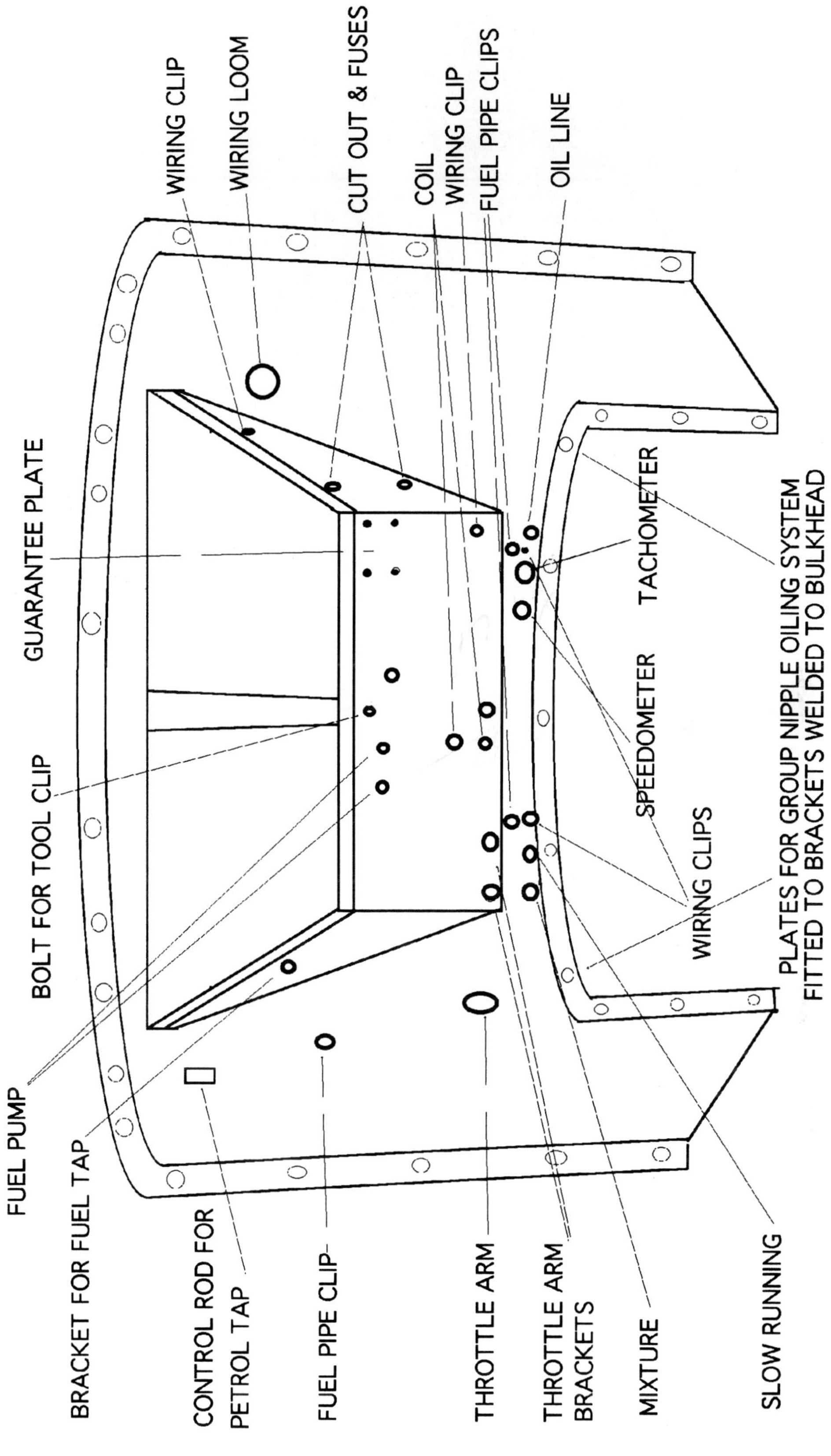

TA BULKHEAD SHOWING HOLES AND POSITION OF MAIN COMPONENTS.
Based on examination of a small number of cars.

TA bulkhead illustrating cut out and fuse box.

Rear view mirror fitted to TA and TB.

Dashboard switch for petrol tap on TA and TB.

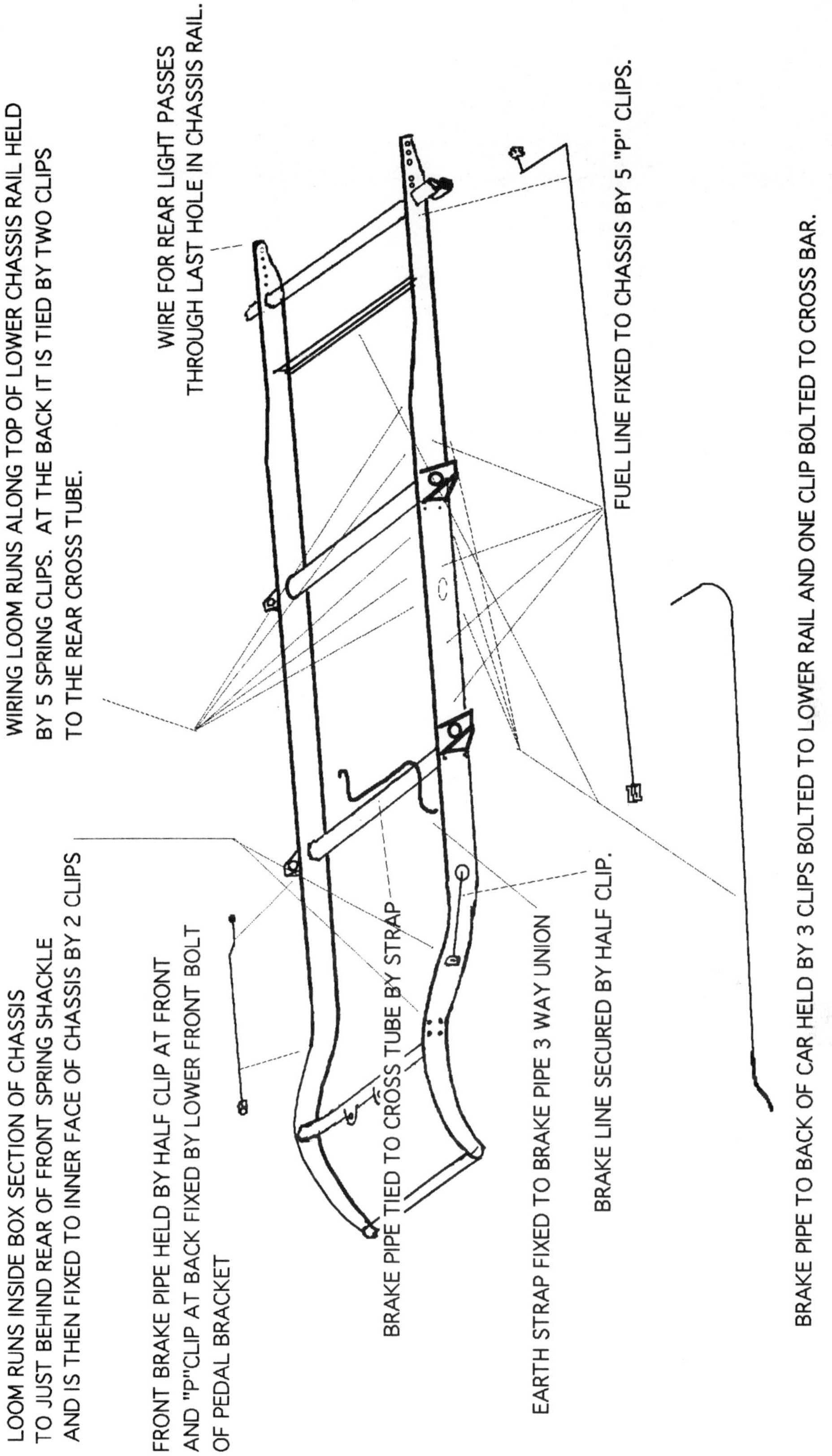

TC CHASSIS DETAILS

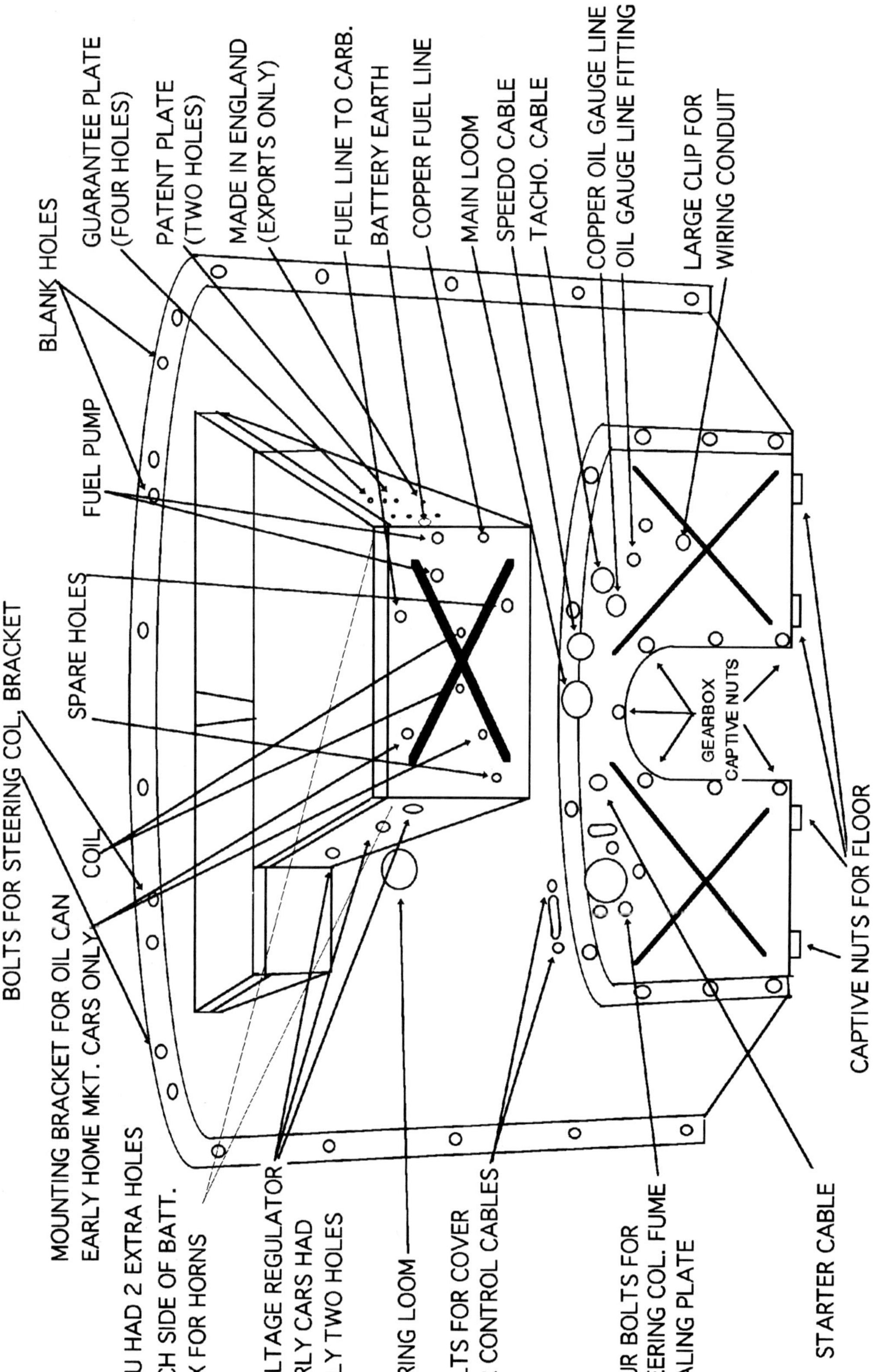

TC BULKHEAD SHOWING POSITION OF HOLES AND MAIN COMPONENTS

The later type of TC regulator. Note the extra hole left over from the earlier style of regulator.

An oil can holder was fitted to the earlier TCs. Some sources say that they were only fitted to home market cars.

The early TC wooden dashboard. This car has a water temperature gauge, passenger's grab handle and Brooklands steering wheel, all of which were popular period accessories.

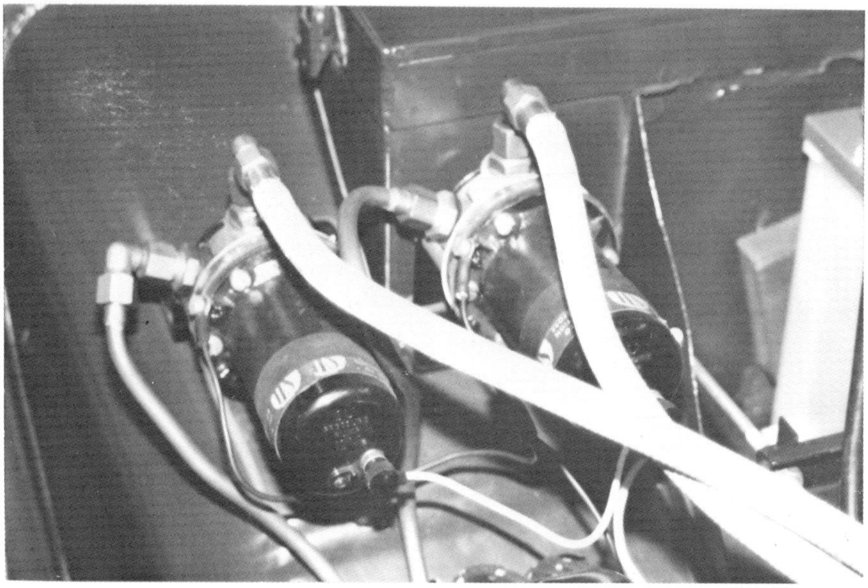

Twin fuel pumps fitted here to the TD Mk ll.

Style of interior mirror fitted to the TD and TF.

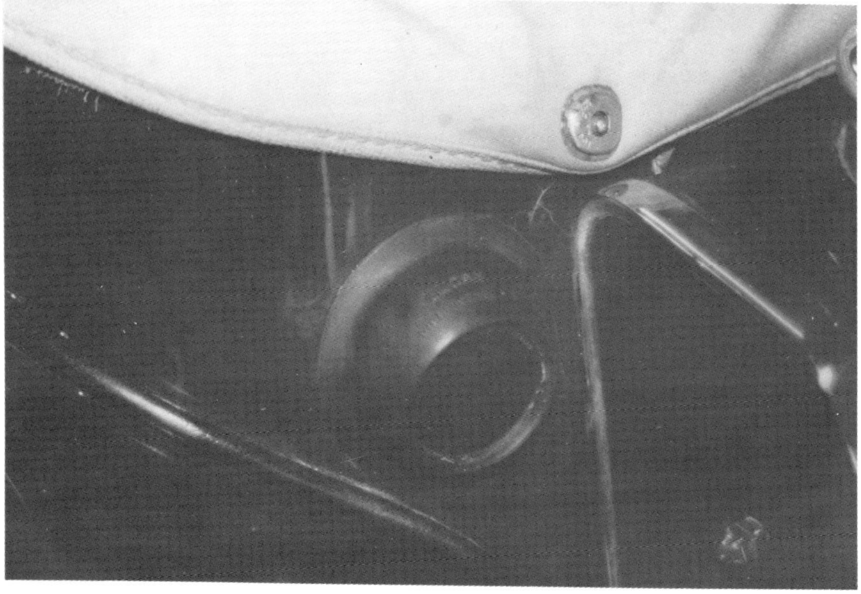

The rear reflector fitted to the TF 1500.

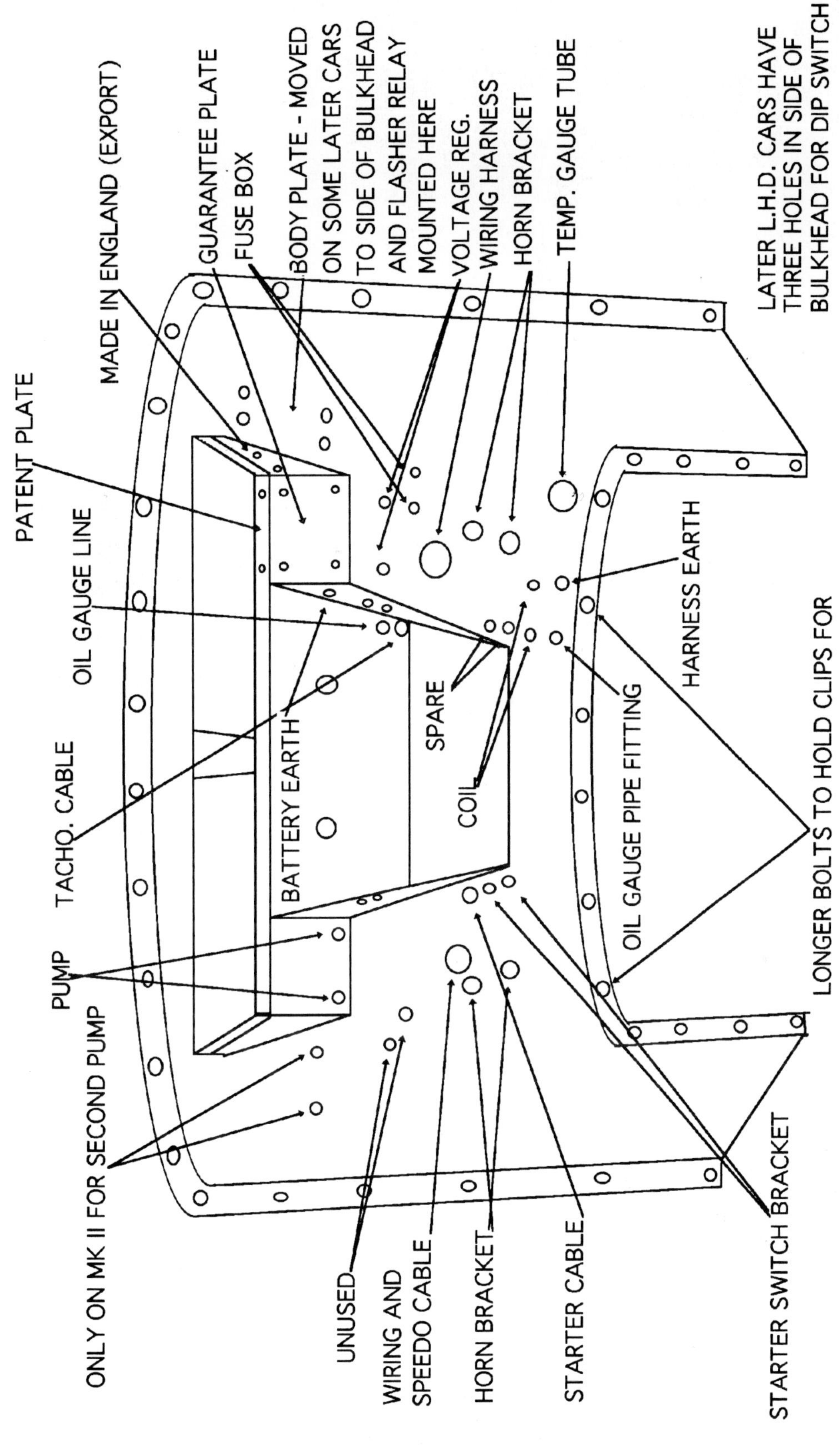

TD BULKHEAD SHOWING POSITION OF HOLES AND MAIN COMPONENTS.

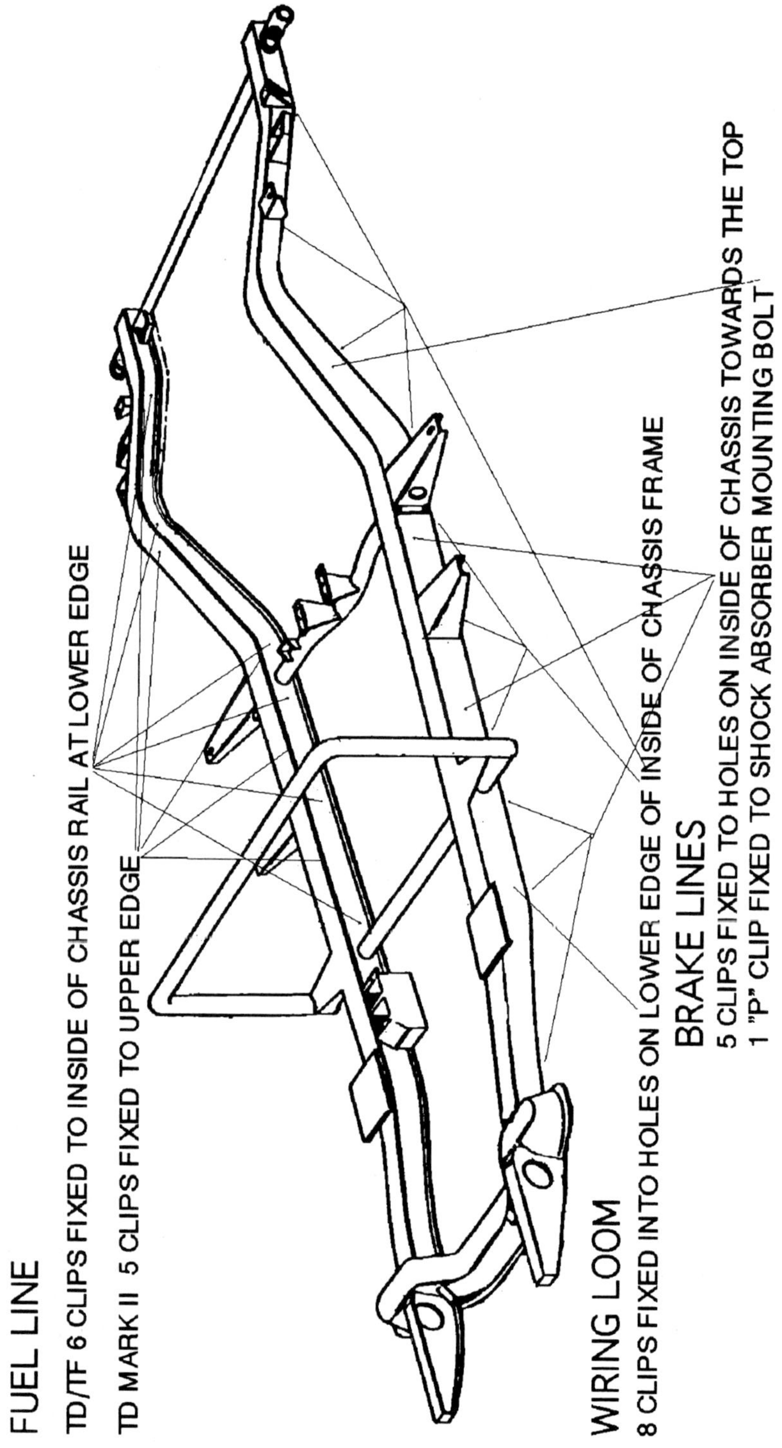

TD/TF CHASSIS SHOWING PIPE AND WIRING CLIPS.

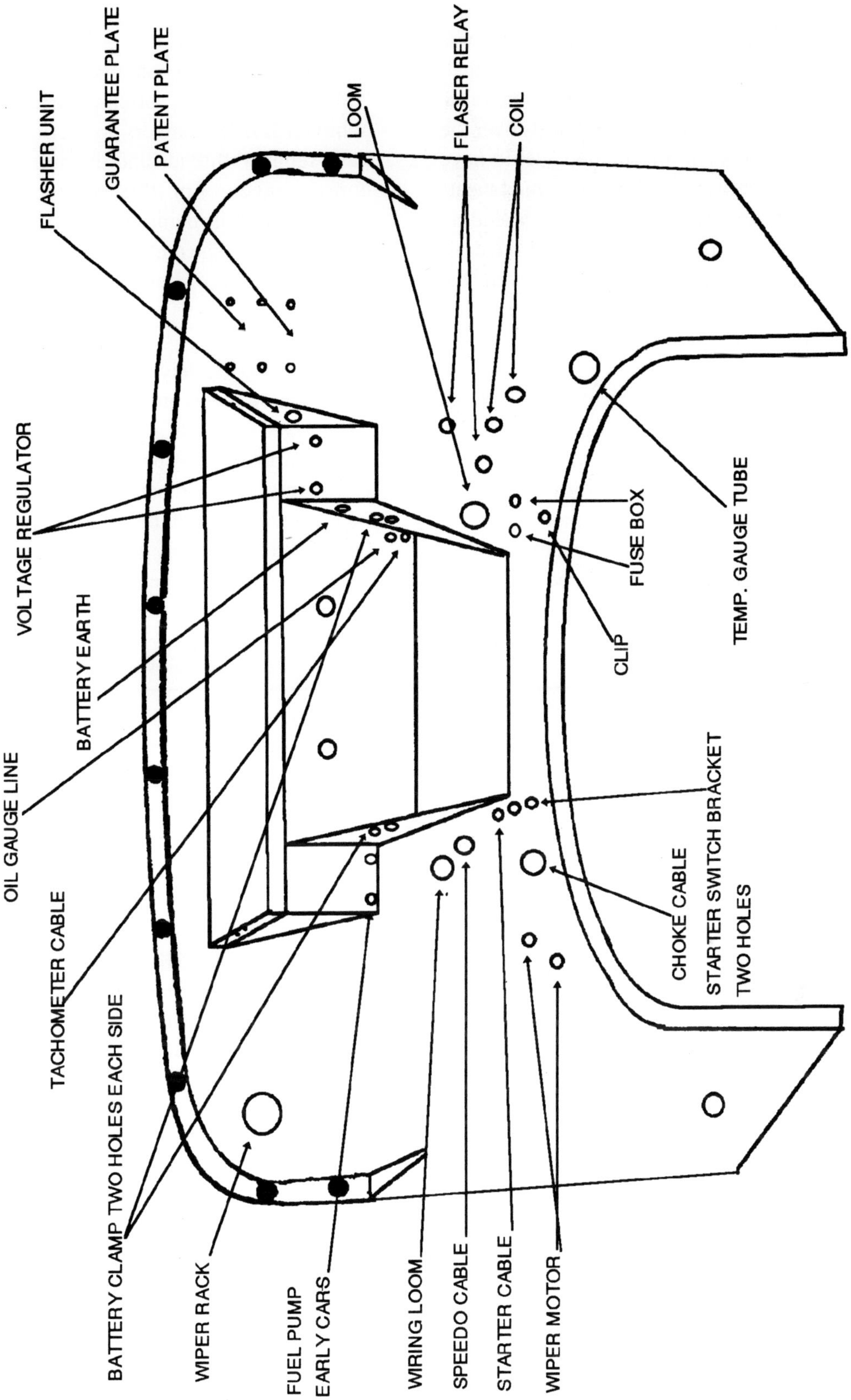

TF BULKHEAD SHOWING POSITION OF HOLES AND MAIN COMPONENTS.

ORIGINAL BOLT SIZES (TAKEN FROM FACTORY DATA SHEETS).

A study of all the fastenings used in the T-type cars could fill a complete book without any certainty that, even then, every variation had been fully explored. Having completed quite a number of rebuilds over the last thirty years, when such information would have been invaluable, I recently obtained photocopied sheets that list the complete specification of the cars, including the sizes of most of the nuts and bolts used. Like most restorers, all I had previously to guide me were the mixed bag of assorted bolts bequeathed by the earlier heavy-handed owners of the cars.

With the standard of many replacement fastenings currently available falling far short of those originally used, it is a good idea to retain as many of the still-serviceable ones as possible. There is a lot of satisfaction in fitting bolts that have the correct manufacturer's markings on their heads and these old bolts can be chemically cleaned, and replated where necessary, just as economically as buying new items. However, I am not advocating re-using any safety critical bolts, or those in the engine. Whilst on the subject of replacement bolts, it is as well to note that some of these now have "odd" sized heads, often metric equivalents, and Whitworth/BSF spanners do not fit. New bolts usually have stamped threads and the nuts are a poorer fit in consequence. For this reason be sure that the replacement bolt being used is strong enough for the intended task. The metric thread bolts supplied for use in the engine are, however, usually of much better quality.

The following lists have been compiled from the original specification sheets but I am not guaranteeing that these are correct for each and every car that left the line. Many variations to specification remain unrecorded. Nor is this by any means an exhaustive list of all the bolts used as some of the photocopied sheets are all but unreadable. I certainly do not want to learn about every car that seems to have been assembled with bolts of different sizes – we have all heard about the Friday afternoon cars!

Position	Description	Diam/thread	Length	Per car
	TA			
CHASSIS FRAME				
Battery carrier stay	Hex. head bolt	5/16BSF	5/8"	6
Third cross tube	Hex. head bolt	5/16BSF	2 1/4"	2
Engine bearers	Hex. head bolt	5/16BSF	1 3/4"	7
	Hex. head bolt	5/16BSF	2"	1
Pedal bracket	Hex. head bolt	5/16BSF	1 3/4"	3
Rear buffers	Hex. head bolt	5/16BSF	3/4"	4
Front rebound rubbers	Hex. head bolt	5/16BSF	5/8"	4
Rear check straps	Hex. head bolt	5/16BSF	5/8"	4
	(Four 3/16" x 5/8" bifurcated rivets fix rubbers to straps)			
Front spring roller bush cover	Hex. head bolt	1/4BSF	1/2"	10
Rear spring front pin	Hex. plain nut	1/2BSF		2
	(with shakeproof lock washer)			
Spring to pad	Hex. head bolt	5/16BSF	3 1/4"	16
	(with Eight 5/16" hex. plain nuts and 5/16" hex. lock nuts)			
ENGINE MOUNTING				
Front engine mounting blocks	Hex. head bolt	5/16BSF	3/4"	4
	(with two 5/16BSF plain nuts on front bolt)			
Rear engine bearer blocks	Hex. head bolt	7/16BSF	1"	4

Position	Description	Diam/thread	Length	Per car
SHOCK ABSORBERS				
Rear shock absorber bracket to frame	Hex. head bolt	5/16BSF	7/8”	10
Front shock absorbers to frame	Hex. head bolt	3/8BSF	1”	4
Rear shock absorbers to bracket	Hex. head bolt	3/8BSF	1 1/8”	4
BRAKE GEAR				
Pipe clips to frame	Hex. head bolt	1/4BSF	1/2”	7
Rear hose bracket to frame	Hex. head bolt	1/4BSF	1/2”	2
Clip to rear axle housing N.S.	Hex. head bolt	5/16BSF	3/4”	1
Three-way union to frame	Hex. head bolt	5/16BSF	3/4”	1
Three-way union to rear axle	Hex. head bolt	3/8BSF	1 1/4”	1
Master cylinder shield	Hex. head bolt	1/4BSF	1/2”	2
Handbrake cable to frame	Hex. head bolt	1/4BSF	5/8”	4
Handbrake outer cable bracket to frame	Hex. head bolt	1/4BSF	1”	2
ELECTRICAL				
Stop lamp switch bkt. to frame	Hex. head bolt	1/4BSF	5/8”	2
Stop lamp switch to bracket	Hex. head bolt	3/16BSF	1/2”	2
Wiring harness clip to frame (1 off)				
Neg. starter cable clips to frame (4 off)				
Battery linking cable clip to frame (1 off)				
Wiring harness to firewall clip (2 off)				
Engine earth wire to frame (1 off)	Hex. head bolt	1/4BSF	3/4”	9
STEERING				
Steering box bracket to frame	Hex. head bolt	5/16BSF	2”	3
Steering box to bracket	Hex. head bolt	3/8BSF	3”	1
FIREWALL/RAMP ASSEMBLY				
Cover plate to ramp	Hex. head bolt	1/4BSF	1/2”	3
Ramp to dash/firewall	Hex. head bolt	5/16BSF	1/2”	5
Dash/firewall to frame bracket	Hex. head bolt	5/16BSF	3/4”	4
Dash to frame bracket and stay tube	Hex. head bolt	5/16BSF	1 1/4”	2
ENGINE COMPARTMENT DETAILS				
Cut-out to firewall	Cheese head machine screw	3/16BSF	1”	2
Guarantee plate (4 off) Patent No. plate (2 off) Toolbox fasteners (6 off)				
	Round head machine screw	5BA	3/8” (with square nuts)	12
Hinge to toolbox lid	Countersunk Woodscrew	5/32 32	1/2”	24
Petrol tap bracket	Hex. head bolt	1/4BSF	1/2”	2
Coil	Hex. head bolt	1/4BSF	1/2”	2
Petrol pump	Hex. head bolt	1/4BSF	1/2”	2
FACIA AND CONTROLS				
Thirtilite and map lamp to facia	Round head woodscrew	No.4	1/2”	2
Thirtilite and map lamp to facia	Round head set-screws	5BA	1 1/4”	2

Position	Description	Diam/thread	Length	Per car
FACIA AND CONTROLS (TA Continued)				
Petrol tap control to facia	Round head woodscrew	No.4	1/2”	2
Tap knob fastening	Dome nut	2BA		
Accelerator pedal bkt. to dash/firewall	Hex. head bolt	1/4BSF	1/2”	2
EXTERNAL FITTINGS				
Bonnet fasteners to bonnet sides	Round head machine screw	2BA	3/8”	8
Rubber protective corners to bonnet sides	Bif. rivets	No.11	7/16”	8
Badge bar bkt. to frame	Hex. head bolt	5/16BSF	1”	4
Clamp screws for brackets	Hex. head bolt	1/4BSF	1”	2
Horn to bracket	Hex. head bolt	3/8BSF	7/8”	1
Wing stay to frame	Hex. head bolt	3/8BSF	2 1/2”	2
Lift-a-Dot 2BA studs - single				5
Lift-a-Dot 2BA studs - double				1
Spare wheel carrier to frame rails	Hex. head bolt	5/16BSF	3/4”	4
Spare wheel carrier centre bracket to frame	Hex. head bolt	5/16BSF	1 1/8”	2
Spare wheel carrier bracket clamp	Hex. head bolt	5/16BSF	1”	1
Spare wheel hub to carrier	Hex. head bolt	3/8BSF	1 3/8”	4
Front splash apron to frame	Hex. head bolt	5/16BSF	3/4”	4
COCKPIT - PROP-SHAFT TUNNEL, FLOOR, ETC.				
Tunnel to frame	Hex. head bolt	5/16BSF	3/4”	4
Fume excluder retainer	Round head woodscrews	No.8	3/4”	6
Starting handle clips	Hex. head bolt	1/4BSF	3/4”	3
CHASSIS LUBRICATION (Later TAs and all TBs)				
Tecalemit lubrication pipes, clips to rails	Round head mach. screw	2BA	5/8”	10
Tecalemit lubrication pipes, clips to firewall	Round head mach. screw	2BA	5/8”	2
Tecalemit lubrication pipes, clip LH frame	Hex. head bolt	1/4BSF	5/8”	1
Tecalemit lubrication pipes, clip RH frame	Hex. head bolt	1/4BSF	7/8”	1
	TB			
As TA except for:				
SHOCK ABSORBERS				
Rear shock absorbers to bracket	Hex. head bolt	3/8BSF	1 1/4”	4
BRAKE GEAR				
Clip to rear axle housing N.S.	Hex. head bolt	1/4BSF	3/4”	1
ENGINE MOUNTING				
Front engine mounting blocks	Hex. head bolt	5/16BSF	3/4”	4
Rear engine bearer blocks	Hex. head bolt	7/16BSF	1 1/4”	2
Rear engine bearer blocks	Hex. head bolt	7/16BSF	2”	2

On TA and TB the front number plate is fixed to the front apron by two 1/4BSF, 1/2" hex. head bolts and the rear plate brackets to the frame and the plate to the brackets by six 1/4BSF, 5/8" hex. head bolts.

Position	Description	Diam/thread	Length	Per car
	TC			
CHASSIS FRAME				
Third cross tube	Hex. head bolt	5/16BSF	2 1/4”	2
Engine mounting bkt.	Hex. head bolt	5/16BSF	1 3/4”	7
Engine mounting bkt.	Hex. head bolt	5/16BSF	2”	1
Pedal bracket to frame	Hex. head bolt	5/16BSF	1 3/4”	2
Pedal bracket to frame	Hex. head bolt	5/16BSF	2”	1
Rear buffers	Hex. head bolt	5/16BSF	3/4”	2
Rear buffers	Hex. head bolt	5/16BSF	1”	2
Front buffers	Hex. head bolt	3/8BSF	5/8”	8
Rear check straps	Hex. head bolt	5/16BSF	5/8”	4
	(Four 3/16 x 5/8 bifurcated rivets fix rubbers to straps)			
Rear spring front pin	Hex. plain nut (with shakeproof lock washer)		1/2BSF	2
Spring to pad	Hex. head bolt	5/16BSF	3 1/4”	16
	(with eight 5/16” hex. plain nuts and 5/16” hex. lock nuts)			
ENGINE MOUNTING				
Front engine mounting blocks	Hex. head bolt	5/16BSF	3/4”	4
		(with two 5/16BSF lock nuts on front bolt)		
Rear engine bearer blocks	Hex. head bolt	7/16BSF	1 3/4”	2
Rear engine bearer blocks	Hex. head bolt	7/16BSF	2”	2
SHOCK ABSORBERS				
Shock absorbers to frame	Hex. head bolt	7/16BSF	2 1/4”	8
Rear shock absorber bracket to frame	Hex. head bolt	5/16BSF	7/8”	1
BRAKE GEAR				
Pipe clips to frame	Hex. head bolt	1/4BSF	1/2”	7
Pipe clips to rear axle pipe sup.	Hex. head bolt	1/4BSF	1/2”	2
Rear hose bracket to frame	Hex. head bolt	1/4BSF	1/2”	2
Clip to rear axle housing N.S.	Hex. head bolt	1/4BSF	3/4”	1
Three-way union to frame	Hex. head bolt	1/4BSF	1 3/8”	1
Three-way union to rear axle	Hex. head bolt	3/8BSF	3/4”	1
Master cylinder shield	Hex. head bolt	1/4BSF	1/2”	2
ELECTRICAL				
Stop lamp switch bkt. to frame	Hex. head bolt	1/4BSF	5/8”	2
Stop lamp switch to bracket	Round head machine screw	3/16BSF	1/2”	2
Wiring harness clip to frame	Hex. head bolt	1/4BSF	5/8”	2
Wiring harness clip to frame	Hex. head bolt	1/4BSF	3/4”	1
Wiring harness to tool box clip	Round head machine screw	2BA	1/2”	1
Engine earth wire to frame	Hex. head bolt	1/4BSF	3/4”	1
STEERING				
Steering box bracket to frame	Hex. head bolt	5/16BSF	2”	3
Steering box to bracket	Hex. head bolt	3/8BSF	3”	1
Steering col. support bkt. to dash	Hex. head bolt	5/16BSF	2.3/4”	2
Steering col. clamp to bkt.	Hex. head bolt	5/16BSF	2 1/4”	1

Position	Description	Diam/thread	Length	Per car
FIREWALL/RAMP ASSEMBLY (TC Continued)				
Ramp to dash/firewall	Hex. head bolt	5/16BSF	1/2"	4
Dash/firewall to frame bracket	Hex. head bolt	5/16BSF	3/4"	4
Dash to frame bracket and stay tube	Hex. head bolt	5/16BSF	1 1/4"	2
Dash bkt. to frame	Hex. head bolt	5/16BSF	5/8"	4
ENGINE COMPARTMENT DETAILS				
Control box to firewall	Round head machine screw	2BA	1 1/4"	2
Guarantee plate	Round head machine screw	5BA	3/8"	4(with sq. nuts)
Patent No. plate	Round head machine screw	5BA	3/8"	2(with sq. nuts)
Clip harness to dash ramp	Hex. head bolt	1/4BSF	5/8"	1
Clips rev. counter and speedo to dash	Hex. head bolt	1/4BSF	1/2"	1
Coil and petrol pump	Hex. head bolt	1/4BSF	1/2"	4
FACIA AND CONTROLS				
Thirtilite and map lamp to facia	Round head woodscrew	No.4	1/2"	2
Thirtilite and map lamp to facia	Round head mach. screw	5BA	1 1/4"	2
Accelerator pedal bkt. to dash/firewall	Hex. head bolt	1/4BSF	1/2"	2
EXTERNAL FITTINGS				
Bonnet fasteners to bonnet sides	Round head machine screw	2BA	3/8"	8
Bonnet fastener bkts. to frame	Hex. head bolt	1/4BSF	1/2"	8
Rubber protective corners to bonnet sides	Bifurcated rivets	No.11	7/16"	8
Bonnet buffers to frame	Hex. head bolt	1/4BSF	1/2"	4
Rear bonnet hinge bkt. to body	Raised head woodscrew	No.10	1"	2
Badge bar bkt. to frame	Hex. head bolt	5/16BSF	1"	4
Clamp screws for brackets	Hex. head bolt	1/4BSF	1"	2
Horn to bracket	Hex. head bolt	3/8BSF	7/8"	1
Wing stay to frame	Hex. head bolt	3/8BSF	2 1/2"	2
Stay tubes to radiator	Hex. head bolt	1/4BSF	7/8"	2
Petrol tank strap bkts to frame	Hex. head bolt	5/16BSF	1 1/2"	2
Petrol tank rubber strips to body	Csk. black jap. woodscrews	No.8	3/4"	9
Petrol pipe clips to frame	Hex. head bolt	2BA	5/8"	5
Petrol pipe clips to dash	Hex. head bolt	2BA	1/2"	2
Spare wheel carrier to frame rails	Hex. head bolt	5/16BSF	3/4"	4
Spare wheel carrier centre bracket to frame	Hex. head bolt	5/16BSF	1 1/8"	2
Spare wheel carrier bracket clamp	Hex. head bolt	5/16BSF	1"	1
Spare wheel hub to carrier	Hex. head bolt	3/8BSF	1 3/8"	4
Front splash apron to frame	Hex. head bolt	5/16BSF	3/4"	4
Rear wings to body	Round head black jap. woodscrews	No. 10	1"	6
Front and rear wings to body	Round head black jap. woodscrews	No. 10	1 1/4"	14
Running boards to wings	Hex. head bolt	1/4BSF	3/4"	6
Running boards to body	Hex. head bolt	1/4BSF	3 1/4"	4
Running boards to body	Hex. head bolt	1/4BSF	3 1/2"	2
Body to dash/firewall	Hex. head bolt	5/16BSF	2 3/4"	16
Body to frame - rear	Hex. head bolt	5/16BSF	2"	2
Body to frame - sides	Hex. head bolt	3/8BSF	1 1/4"	2
Body to frame sides	Hex. head bolt	3/8BSF	1"	2

Position	Description	Diam/thread	Length	Per car
COCKPIT - PROP-SHAFT TUNNEL, FLOOR, ETC.				
Tunnel to frame	Hex. head bolt	5/16BSF	3/4"	4
Tunnel to rear board	Hex. head bolt	1/4BSF	1 1/8"	5
Starting handle clips	Hex. head bolt	1/4BSF	3/4"	3
Gearbox cover fixing	Csk. metal thread screws	1/4BSF	1/2"	1
Gearbox cover fixing	Csk. metal thread screws	1/4BSF	5/8"	5
Gearbox cover fixing	Round head woodscrews	No. 8	3/4"	12
G/box cover and speedo cable clip	Csk. machine screw	1/4BSF	1 1/4"	1
Floorboards to body rails	Csk. machine screws	1/4BSF	1 1/4"	6
Floorboards to body rails	Csk. machine screws (pointed)	1/4BSF	1"	4
Sliding seat fittings	Csk. steel woodscrews	No. 8	3/4"	12
The body identification plate is fixed to the body with four 1/2" escutcheon pins.				
	TD/TF			
CHASSIS FRAME AND SUSPENSION				
Scuttle bracing bracket to frame	Hex. head bolt	5/16BSF	7/8"	8
Scuttle bracing bracket to firewall/ramp	Hex. head bolt	5/16BSF	3/4"	10
Bracket to stay tube	Hex. head bolt	3/8BSF	3 1/4"	2
Stay tube to chassis rail	Hex. head bolt	3/8BSF	1 1/8"	2
Spring pan to wishbone	Hex. head bolt	5/16BSF	7/8"	4
Spring pan to wishbone	Hex. head bolt	5/16BSF	1"	4
Wishbone tie bolt	Hex. head bolt	3/8BSF	2 3/4"	2
Wishbone to link bolt	Hex. head bolt	1/2BSF	4 1/8"	2
Lockheed dampers:				
Damper to chassis	Hex. head bolt	3/8BSF	1 1/8"	4
Damper to chassis	Hex. head bolt	3/8BSF	1 1/4	3
Damper to chassis	Hex. head bolt	3/8BSF	1 7/8	1
Girling Dampers;				
Damper to chassis	Hex. head bolt	3/8BSF	1 3/4"	4
Damper to chassis	Hex. head bolt	3/8BSF	2 1/8"	4
Wishbone pivots to chassis	Hex. head bolt	3/8BSF	1 1/8"	8
Rear axle check strap to frame	Hex. head bolt	3/8BSF	3/4"	4
ENGINE MOUNTING				
Engine control bkt. to frame	Hex. head bolt	3/8BSF	3/4"	2
Rear engine bearer bkt. to frame	Hex. head bolt	3/8BSF	3/4"	2
Front engine mounting to engine	Hex. head bolt	3/8BSF	3/4"	2
BRAKE GEAR				
Front brake back plate to steering knuckle	Hex. head bolt	3/8BSF	7/8"	8
Three-way pipe union to frame	Hex. head bolt	5/16BSF	3/4"	1
Three-way pipe union to diff.	Hex. head bolt	5/16BSF	5/8"	1
Master cylinder to frame	Hex. head bolt	3/8BSF	3"	2
ELECTRICAL				
Regulator box to firewall	Rd. hd. Phillips screw	2BA	1 1/8"	2
Dip switch to bkt. (f/wall mount)	Hex. head bolt	2BA	3/8"	2

Position	Description	Diam/thread	Length	Per car
STEERING (TD/TF Continued)				
Steering column support bkt. to frame	Hex. head bolt	5/16BSF	7/8"	2
Steering column clamp	Hex. head bolt	5/16BSF	2 3/8"	1
Steering col. fume seal ret. to ramp	Hex. head bolt	5/16BSF	7/8"	3
Steering rack to frame	Hex. head bolt	5/16BSF	1 1/4"	4
EXTERNAL FITTINGS TD				
Firewall to body tub	Hex. head bolt	5/16BSF	2 1/2"	16
Firewall to body tub	Hex. head bolt	5/16BSF	2 3/4	4
Running boards to body	Hex. head bolt	1/4BSF	3 1/4"	2
Running boards to body	Hex. head bolt	1/4BSF	3"	4
Running boards to front wings	Hex. head bolt	5/16BSF	1"	6
Front wings to chassis	Hex. head bolt	5/16BSF	3/4"	12
Rear wings to body	Hex. head bolt	5/16BSF	3/4"	8
Rear wings to body	Hex. head bolt	5/16BSF	1 1/4"	2
Fr. and rear wings to body	Rd. hd. Phillips S/T screw	No. 8	1/2"	6
Front apron to wing	Hex. head bolt	5/16BSF	3/4"	2
Front apron to wing	Rd. hd. Phillips mach. screw	5/16BSF	3/4"	4
Rear valance to frame	Rd. hd. Phillips mach. screw	5/16BSP	3/4"	2
Bonnet fastener bkts to wings	Hex. head bolt	1/4BSF	1/2"	8
Spare wheel carrier to body	Hex. head bolt (cad. plated)	5/16BSF	2"	4
EXTERNAL FITTINGS TF				
Front wings to sides	Hex. head bolt	5/16BSF	7/8"	10
Front wings to sides	Hex. head bolt	5/16BSF	1"	4
Running boards to body	Hex. head bolt	1/4BSF	3 1/4"	2
Running boards to body	Hex. head bolt	1/4BSF	3"	4
Front fairing to wings	Hex. head bolt	1/4BSF	3/4"	4
Front fairing to rad. case	Rd. hd. Phillips mach. screw	1/4BSF	1"	2
Front wings to running boards	Hex. head bolt	5/16BSF	1"	6
Front wings to frame	Hex. head bolt	5/16BSF	1"	4
Wings to body	Hex. head bolt	1/4BSF	1 1/8"	2
Wings to body	Hex. head bolt	1/4BSF	3/4"	14
Rear fairing to frame	Rd. hd. Phillips mach. screw	5/16BSF	3/4"	2
Bonnet fstnr. catch assy. to body	Raised head mach. screw	2BA	5/8"	8
Fastener keep plates to bonnet	Hex. head set screw	1/4BSF	3/8"	8
Spare wheel support bracket to outer frame	Hex. head bolt	7/16BSF	4"	2
COCKPIT - PROP-SHAFT, TUNNEL, FLOOR, ETC.				
Brake/clutch cover plate to bracket	Hex. head bolt	1/4BSF	1/2"	3
Fume excluder retainer	Round head woodscrew	No.8	1/2"	6
Floorboards to body rail	Csk. zinc plated Phillips woodscrew	No. 8	1 1/4"	8
Floorboard to p/sft. tunnel support	Csk. pointed mach. screw	1/4BSF	1 1/4"	2
Floorboard to body & tunnel support	Csk. pointed mach. screw	1/4BSF	1 1/8"	11
Gearbox and p/sft tunnel to floor	Csk. pointed mach. screw	1/4BSF	1 1/8	5
G/box & p/sft tunnel to floor & ramp	Csk. pointed mach. screw	1/4BSF	3/4"	8

(All above floor fixing machine screws are zinc plated and sit in 1/4" cup washers. Specification sheets list rubberised anti-drum felt over gearbox cover.)

APPENDIX

TD RADIO INSTALLATION

These instructions were issued to dealers installing the optional radio in new TDs. It is rare now to find an original 1950s valve car radio in full working order but there are a number of specialists who will restore old equipment to health, or even fit modern components within the old casing to give the radio much improved performance.

INSTALLATION INSTRUCTIONS FOR

"HIS MASTER'S VOICE" AUTOMOBILE RADIO

MODELS 4100-4102-4200-4202 SERIES

SMITHS RADIOMOBILE

EMITRON CAR RADIO

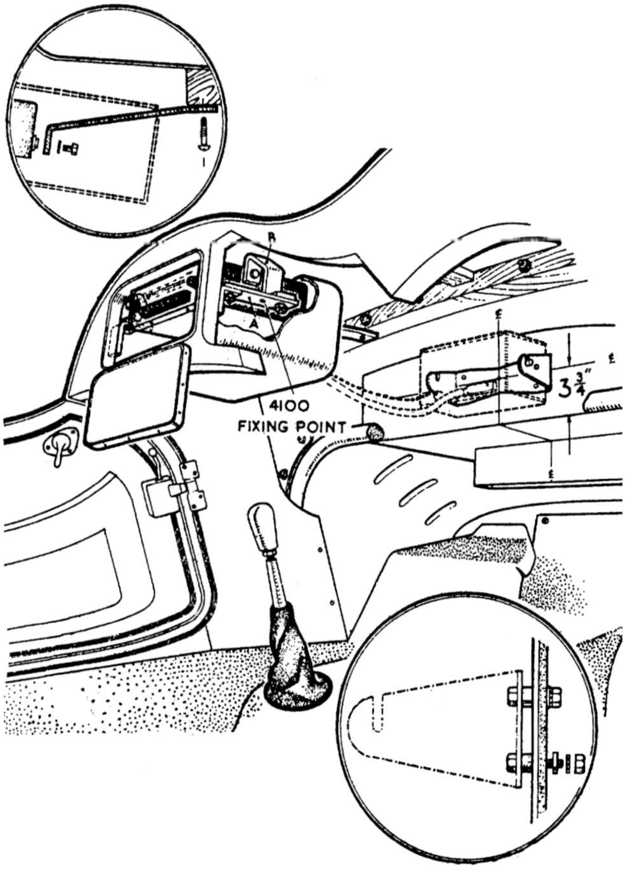

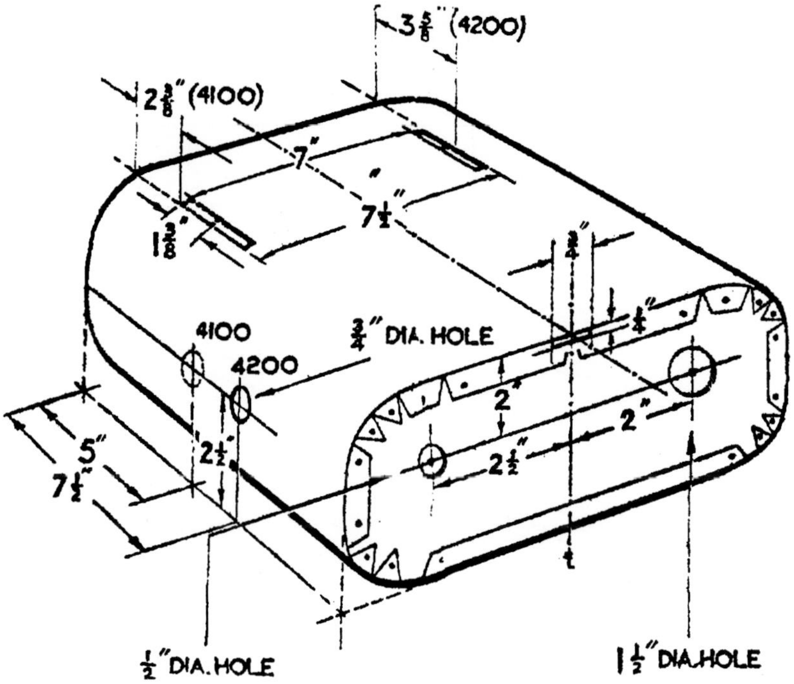
3 5/8" (4200)
2 3/8" (4100)
7"
7 1/2"
1 3/8"
3/4"
1/4"
4100
4200
3/4" DIA. HOLE
2"
2"
2 1/2"
5"
7 1/2"
2 1/2"
1/2" DIA. HOLE
1 1/2" DIA.HOLE

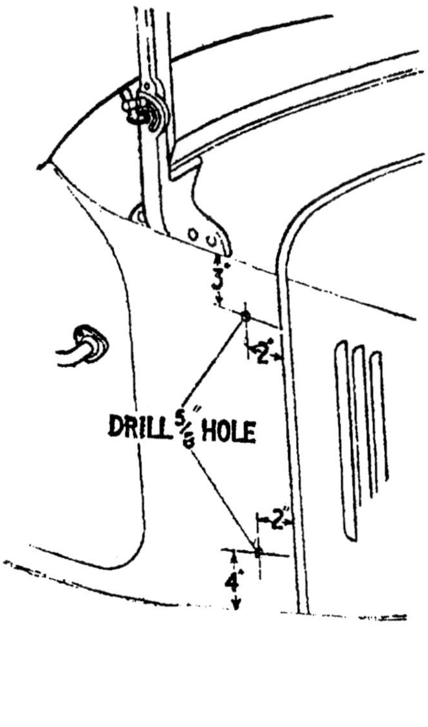
3"
2"
DRILL 5/8" HOLE
2"
4"

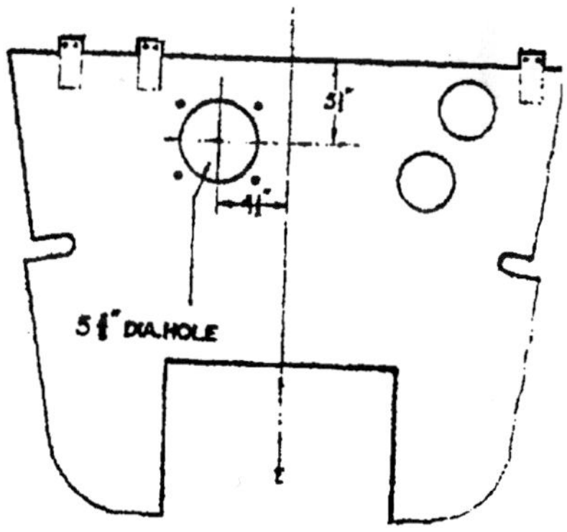

Service Information Sheet No. 67

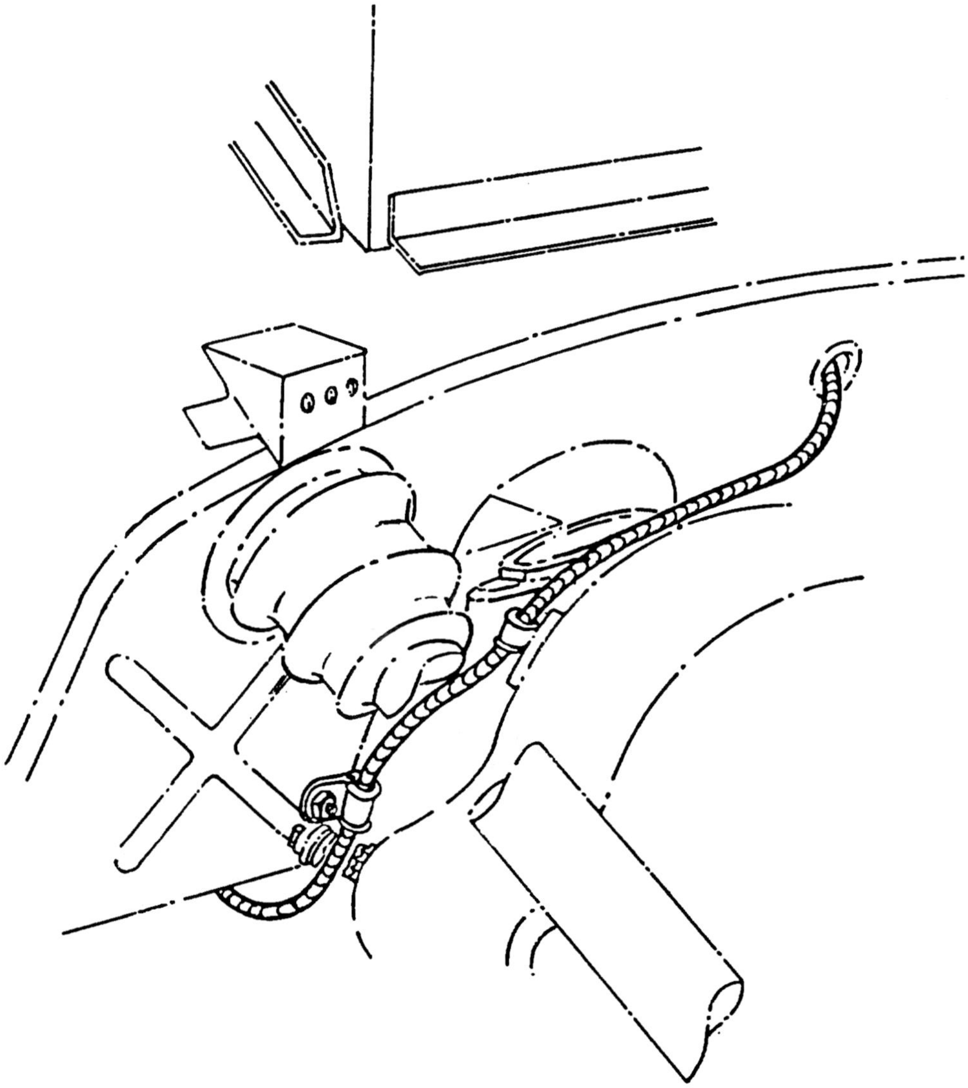

View showing speedometer flexible cable run between foot ramp and clutch housing.

Date of Issue : August, 1950

Service Information Sheet No. 80

EXCESSIVE OIL CONSUMPTION

Some early cars of the " TD " Series have been observed to have a heavy oil consumption.

In some cases this has been found to be due to oil passing from the valve cover into the air cleaner and being consumed in the engine.

This can be prevented by inserting a restrictor or washer in the air cleaner engine breather pipe as indicated in the sketch below.

Later production engines have this modification incorporated as standard.

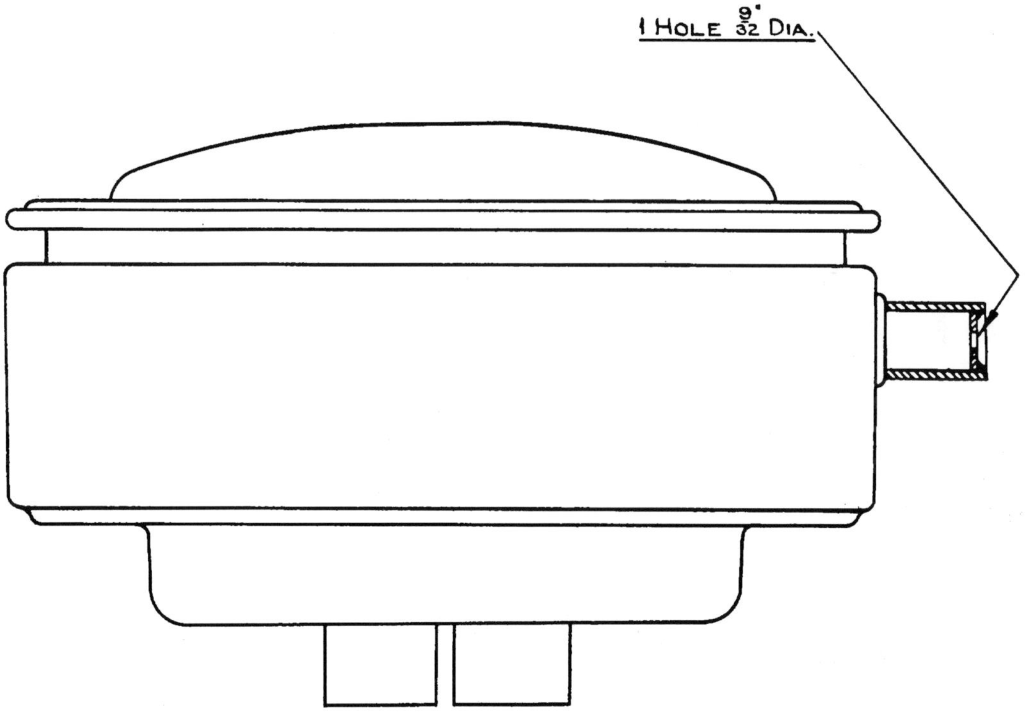

Date of Issue: August, 1950

Service Information Sheet No. 81

GEARBOX SYNCHROMESH HUB BALLS

Cases have been encountered where the synchromesh hub balls have dropped out of their cage into the bottom of the gearbox, necessitating the dismantling and reassembly of the complete gearbox.

To obviate this, extreme care must be taken when removing and refitting the gear change lever not to withdraw the third speed gear selector shaft past the first stop, as this will cause the two components of the synchromesh mechanism to slide apart, releasing the balls and springs, which will drop into the box.

The third speed selector shaft is provided with two stops—one of the spring-loaded ball type, the other consisting of a boss on the inside face of the gear-lever housing. When the housing is removed there is no longer a rear stop to arrest the movement of the third speed selector shaft, which is liable to be withdrawn beyond its normal travel.

Reference to the accompanying illustration of the gearbox will make this clear.

On later models a longer selector shaft is fitted which has a retaining circlip at its forward end to prevent it from being withdrawn too far.

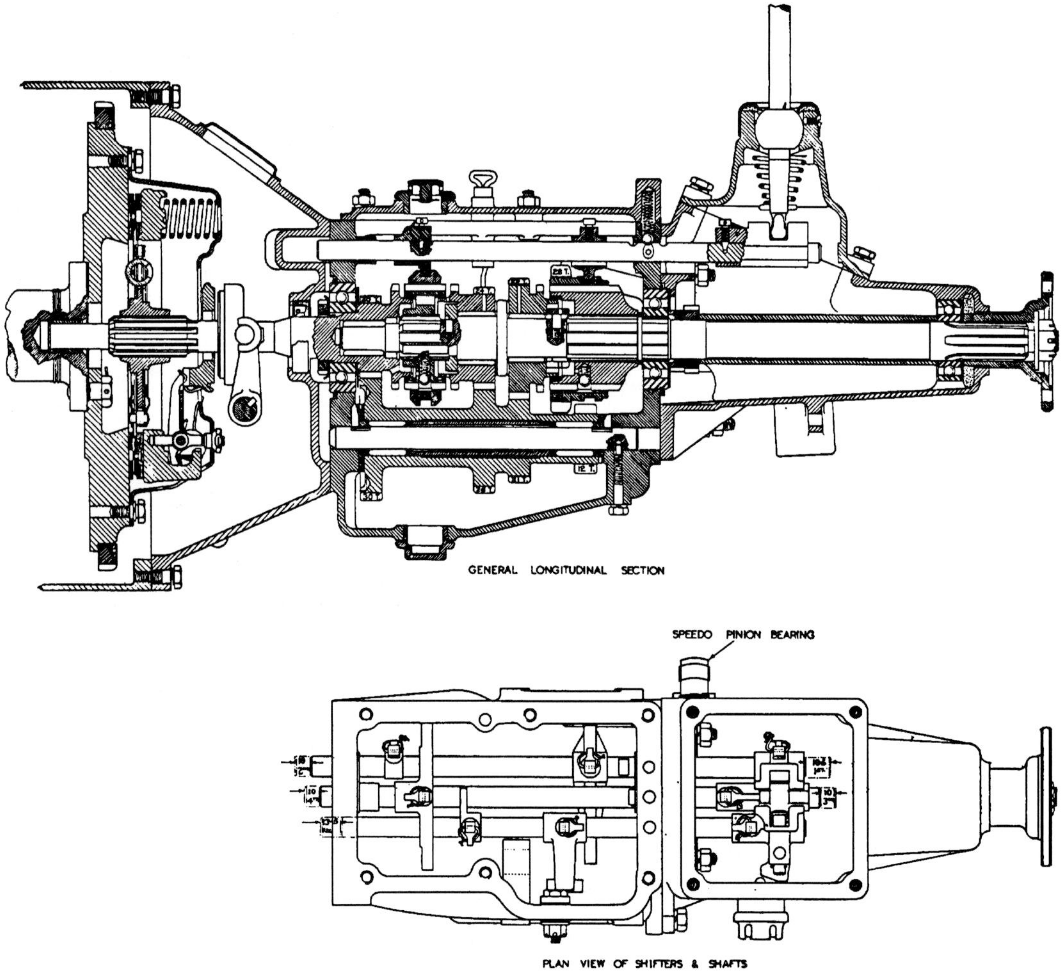

Sectional view of the gearbox.

LUCAS

ELECTRICAL EQUIPMENT

M.G. T.C. MIDGET

EXPORT—U.S.A.

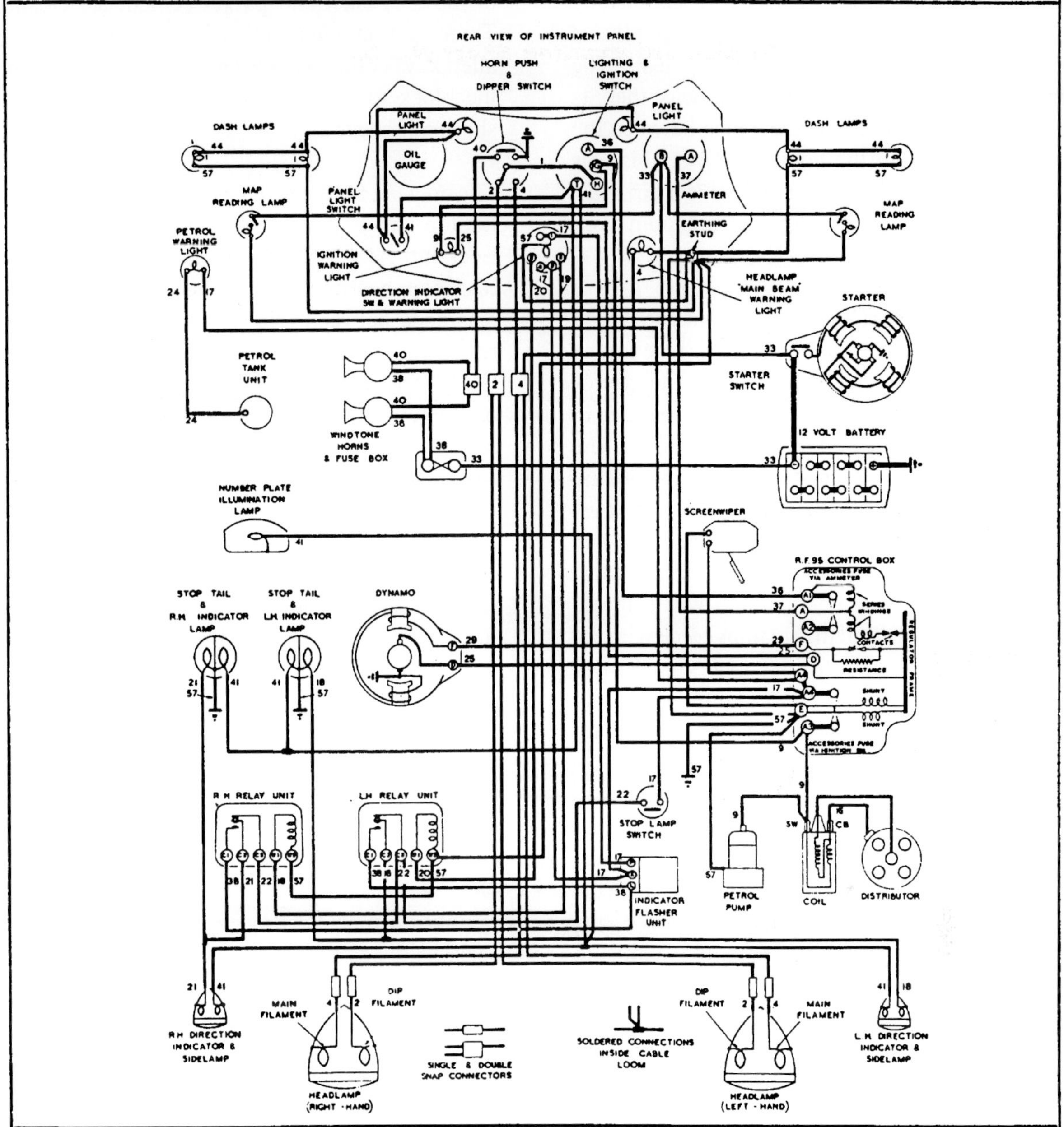

KEY TO CABLE COLOURS

1 BLUE
2 BLUE with RED
3 BLUE with YELLOW
4 BLUE with WHITE
5 BLUE with GREEN
6 BLUE with PURPLE
7 BLUE with BROWN
8 BLUE with BLACK
9 WHITE
10 WHITE with RED
11 WHITE with YELLOW
12 WHITE with BLUE
13 WHITE with GREEN
14 WHITE with PURPLE
15 WHITE with BROWN
16 WHITE with BLACK
17 GREEN
18 GREEN with RED
19 GREEN with YELLOW
20 GREEN with BLUE
21 GREEN with WHITE
22 GREEN with PURPLE
23 GREEN with BROWN
24 GREEN with BLACK
25 YELLOW
26 YELLOW with RED
27 YELLOW with BLUE
28 YELLOW with WHITE
29 YELLOW with GREEN
30 YELLOW with PURPLE
31 YELLOW with BROWN
32 YELLOW with BLACK
33 BROWN
34 BROWN with RED
35 BROWN with YELLOW
36 BROWN with BLUE
37 BROWN with WHITE
38 BROWN with GREEN
39 BROWN with PURPLE
40 BROWN with BLACK
41 RED
42 RED with YELLOW
43 RED with BLUE
44 RED with WHITE
45 RED with GREEN
46 RED with PURPLE
47 RED with BROWN
48 RED with BLACK
49 PURPLE
50 PURPLE with RED
51 PURPLE with YELLOW
52 PURPLE with BLUE
53 PURPLE with WHITE
54 PURPLE with GREEN
55 PURPLE with BROWN
56 PURPLE with BLACK
57 BLACK
58 BLACK with RED
59 BLACK with YELLOW
60 BLACK with BLUE
61 BLACK with WHITE
62 BLACK with GREEN
63 BLACK with PURPLE
64 BLACK with BROWN

WIRING DIAGRAM

No. W75841

12-VOLT

MODERN OIL SEALS FOR THE XPAG ENGINE

The propensity of the XPAG engine to leak oil has been covered in some detail in the engine section of this book but modern technology, and a deal of ingenuity by enthusiasts like Al. Moss, have resulted in rear oil seal kits being available that use a modern lip seal. The trouble with the original arrangement was that its efficiency depended on a tight clearance between the scroll on the crank and the housing. Once this was worn, usually in an engine where excessive main bearing wear had occurred, only major machine room work would remedy the defects.

The modern replacement seal uses the outside rim of the rear crankshaft flange as the bearing surface for the lip of a large seal. This seal is contained within a two-part housing that is fixed to the rear face of the block in place of the old oil thrower, which is discarded. The seals are usually totally successful, early failure being traced to poor installation – except in some rare cases that I will cover later. As the seal is running on the rim of the rear flange, the condition of this is of the utmost importance, any slightest imperfection will ruin the new seal in short order. Full instructions come with the seal kits, which are available from suppliers like Brown and Gammons, and these MUST be carefully read and adhered to. To give some idea how the seal is installed, a brief summary of the instructions follows.

To fit the seal the engine must be removed from the car and dismantled at least to the point where the flywheel is separated from crankshaft and they are both out of the engine. As already explained, the outer rim of the flange must be examined and given to a machinist to face and polish if any defects are found. As an alternative, the thin sleeve mentioned later can be fitted. The three bolts holding the oil thrower are removed and the thrower discarded, the pins can be left in situ; all traces of the oil gasket and gasket cement are cleaned off. Three new countersunk screws (and an appropriate allen key) come in the kit and these should be tried in the block to make sure that they are not overlong; shorten if needed.

The housing for the new gasket is separated in two and the top half is fixed to the block with screws, using Locktite sealer in place of the old gasket. Lightly fit the rear main bearing cap and check to see that there is a clearance between this and the top half of the seal retainer. If not it will be necessary to face off the OUTSIDE ONLY of the bearing cap to give a clearance of .003". The second part of the seal is then offered up to ensure that this too is clear of the cap, if not grind sufficient off the rear of the cap to provide clearance.

A 3/16" hole is then drilled through the bearing cap, aligning with the one in the lower half of the seal retainer. This is to allow the oil to drain into the trough in the main cap and thence back into the sump. All the drill swarf must be carefully removed. Leaving the main bearing cap in place, remove the lower, thinner, section of the seal retainer and apply beads of sealer to the two grooves. Unite the two halves of the retainer and secure the two socket screws. Ensure the drilled hole has not been blocked by sealer.

Once the sealer has cured, it is best left overnight, remove the main bearing cap and the lower half of the seal retainer as a unit taking care not to separate them.

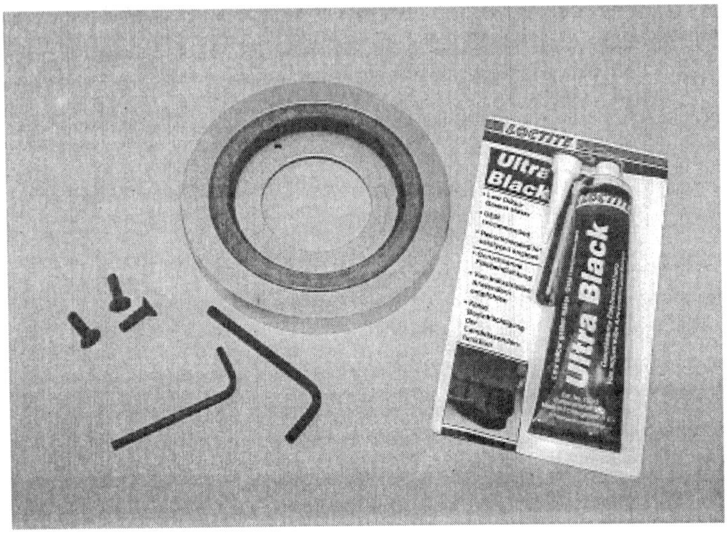

The rear oil seal kit comprises a two-part retainer for the large lip seal, the correct sealer and fixing bolts with appropriate allen keys.

The crankshaft can now be installed. The upper bearing shells are inserted in their housings and lubricated, the new oil seal is carefully fitted to the rear flange of the crankshaft with the lip of the seal facing towards the front of the engine after lightly greasing the flange, and the mating surfaces of the seal housing are given a thin coat of sealer. The crank is then lowered into position, ensuring the seal fits into the housing. The bearing shells are fitted into the caps and lubricated, the front and centre caps are fitted and the nuts secured finger tight before the rear housing, complete with the lower half of the housing is fitted. The bolts are then tightened to the correct torque setting and the free rotation of the crank checked.

The flywheel can then be fitted, using some sealer on the dowel pins and bolts. These must be checked to ensure that they do not foul the seal retainer when tightened, if they do then they must be shortened by grinding or filing. New bolts should be used as a matter of course and tightened to the correct torque setting. Leaks here should now be eliminated!

Although the vast majority of seal kits have been installed without problem there have been a few isolated cases where there has either been insufficient width of surface on the flange of the crank – usually on older crankshafts – or where the rear face of the block was machined by the factory to a greater extent than usual. In these cases the seal would be allowed to ride too far forward on the rim of the flange and leaks could occur. The solution advocated by MG suppliers, Brown and Gammons, is to use something called "Speedi-sleeve". This is an ultra-thin stainless steel tube that fits over the flange, held in place with Loctite, that provides an ideal bearing surface for the seal. They are able to supply the correct-sized tube in kit form.

The rear camshaft plug, it looks like a core plug, is prone to leak unless carefully fitted and leaks here are severe as the camshaft bearing is lubricated at full pressure. A sealing solution like JB Weld should be used when the plug is installed and care taken to ensure the plug is "spread" correctly to make sure it cannot work loose.

Having cured leaks from the rear of the engine, what about those from the front of the crankshaft, around the seal in the timing chain cover/sump? On page 81 there is a drawing of the factory seal arrangement but with modern seals of the correct size easily available a much better solution is to use one of these. Once again it is imperative that the surface of the pulley that the seal will utilise is smooth, to avoid damaging it. A rubber cased lip seal of dimensions 35 x 47 x 7 mm will fit in the existing housing and I have successfully used one this size. With the housing thoroughly clean, the seal can be fitted as the sump is installed, using a silicone sealer round the oil seal to close any small gaps. Some gasket sets include such a seal in place of the now banned asbestos sealing strips.

A neater and better engineered solution is to have the timing chain cover and sump machined to take a 50 mm diameter, metal-cased seal – one such installation I have seen allows the seal to be prised out and replaced, if required, without removing the sump and timing chain cover from the engine. However, the simpler system seems to work quite well enough for most purposes and requires no non-standard modifications.

The rear crankshaft oil seal in place on an XPAG engine. As mentioned in the text, care must be taken to ensure that the seal lip is actually bearing on a wide enough section of the rear crankshaft flange. If not then the installation of a piece of "Speedi-sleeve" to provide a perfect surface for the seal is recommended.

GENERAL DATA

TA

Bore	63.5 mm
Stroke	102 mm
Capacity	1292cc
Horse-power (RAC rating)	10
Sump capacity	(approx.) 1 3/8 Galls.
Firing order	1, 3, 4, 2
Sparking plugs	Champion L.10 (14 mm)
Capacity of water system	1 7/8 galls.
Capacity of gearbox	2 pints
Capacity of rear axle	2 pints
Capacity of petrol tank	(approx.) 13 1/2 galls.

Dimensions (Two-seater):-	
Overall length	11' 7 1/2"
Overall width	4' 8"
Overall height (hood up)	4' 5"
Wheelbase	7' 10"
Track	3' 9"
Weight (Two-seater)	15 3/4 cwt.

Tyre pressures (front tyres)	24 lb.
Tyre pressures (rear tyres)	26 lb.
Front wheel toe-in	3/16"

Gear Ratios

Overall ratios *Ceasing Eng. No. MPJG.683*		mph at 1000 rpm	Overall ratios *Commencing Eng. No MPJG.684*		mph at 1000 rpm
Top	4.875 to 1	16.67	Top	4.875 to 1	16.67
3rd	6.92 to 1	11.73	3rd	6.44 to 1	12.62
2nd	10.73 to 1	7.58	2nd	9.95 to 1	8.17
1st	18.10 to 1	4.49	1st	16.84 to 1	4.83

Rear Axle Ratio:- 4.875 to 1

Valve Timing:-

Inlet opens 11° before T.D.C.
Closes 59° after B.D.C.
Exhaust opens 56° before B.D.C.
Closes 24° after T.D.C.

Tappet clearances	.015" hot.
Contact breaker points	.010/.012".
Ignition timing	T.D.C.

GENERAL DATA

TB

Bore	66.5 mm
Stroke	90 mm
Capacity	1250 cc
Horse-power (RAC rating)	10.97
Sump capacity	(approx.) 1 3/8 galls.
Firing order	1, 3, 4, 2
Sparking plugs	Champion L.10.S (14 mm)
Capacity of water system	1 3/4 galls.
Capacity of gearbox	1 1/2 pints
Capacity of rear axle	2 pints
Capacity of petrol tank	(approx.) 13 1/2 galls.

Dimensions (two-seater):-

Overall length	11' 7 1/2"
Overall width	4' 8"
Overall height (hood up)	4' 5"
Wheelbase	7' 10"
Track	3' 9"
Weight (two-seater)	15 1/2 cwt.

Tyre pressures (front tyres)	24 lb.
Tyre pressures (rear tyres)	26 lb.
Front wheel toe-in	3/16"

Gear Ratios:-

Overall ratios		mph per 1000 rpm
Top	5.125 to 1	15.84
3rd	6.93 to 1	11.74
2nd	10.00 to 1	8.13
1st	17.32 to 1	4.68

Rear Axle Ratio:- 5.125 to 1

Valve Timing:-

Inlet opens 11° before T.D.C.
Closes 57° after B.D.C.
Exhaust opens 52° before B.D.C.
Closes 24° after T.D.C.

Tappet clearances	.019" hot
Contact breaker points	.010/.012"
Ignition timing	T.D.C.

GENERAL DATA

TC

Bore	66.5 mm
Stroke	90 mm
Capacity	1250 cc
Horse-power (RAC rating)	10.97
Sump capacity	1 1/8 galls
Firing order	1, 3, 4, 2
Sparking plugs	Champion L.10.S (14 mm)
Capacity of water system	1 3/4 galls.
Capacity of gearbox	1 1/2 pints
Capacity of rear axle	2 pints
Capacity of petrol tank	13 1/2 galls.

Dimensions:-

Overall length	11' 7 1/2"
Overall width	4' 8"
Overall height (hood up)	4' 5"
Wheelbase	7' 10"
Track	3' 9"
Weight	15 1/2 cwt.

Tyre pressures (front tyres)	24 lb.
Tyre pressures (rear tyres)	26 lb.
Front wheel toe-in	3/16"

Gear Ratios:-

	Overall ratios	Mph per 1000 rpm
Top	5.125 to 1	15.84
3rd	6.93 to 1	11.74
2nd	10.00 to 1	8.13
1st	17.32 to 1	4.68

Rear Axle Ratio :- 5.125 to 1

Valve timing:-

Inlet opens 11° before T.D.C.
Closes 57° after B.D.C.
Exhaust opens 52° before B.D.C.
Closes 24° after T.D.C.

Tappet clearances	.019" hot
Contact breaker points	.010/.012"
Ignition timing T.D.C.	

GENERAL DATA

TD

Engine type	XPAG/TD up to Engine No. 9407 XPAG/TD2 from Engine No. 9408
Number of cylinders	4
Capacity	1250 cc (76 1/4 cu. in.)
BHP	54.4 at 5200 rpm
Bore	66.5 mm (2.618")
Stroke	90 mm (3.543")
RAC rating	10.97 hp
Compression ratio	7.25 to 1
System of cooling	Thermo-siphon - pump and fan assisted
Radiator hose top	Length 4" (10.16 cm) Diameter 2 3/16" (5.56 cm)
Radiator hose bottom	Length 2 1/4" (5.71 cm) Diameter 1 1/8" (2.86 cm)
First oversize bore	+.020" (.50 mm) Actual bore 2.6381" (67 mm)
Second oversize bore	+.040" (1.00 mm) Actual bore 2.6581" (67.5 mm)
Firing order	1, 3, 4, 2
Piston clearance at pressure face below oil ring	.0021" to .0029" (.056 mm to .071 mm)
Ring gap	.006" to .010" (.152 mm to .254 mm)
Number of compression rings	Two
Width of compression rings	2.25 mm (.0885")
Number of oil rings	One
Width of oil ring	4.0 mm (.1575")
Oil pressure relief valve operates	50 to 70 lb per sq in (3.5 to 5.0 kg/cm^2)
Oil pressure (normal)	40 to 45 lb per sq in (2.8 to 3.2 kg/cm^2)
Gudgeon pin type	Clamped
Gudgeon pin diameter	.7087 (+.0004"/+.0006") 18mm (+.010mm/+.015mm)
Fit in piston	Double thumb press
Fit in connecting rod	Clamped
Crankpin diameter (standard)	1.772" (45 mm)
Crankpin minimum diameter for regrind	1.722" (43.75 mm)
Connecting rod - length between centres	7.008" (178 mm)
Connecting rod - type of bearing	Shimless, steel-backed, white-metal-lined
Connecting rod - side clearance	.004" to .006" (.10 mm to .15 mm)
Connecting rod - diametrical clearance	.0005" to .002" (.011 mm to .056 mm)
Number of crankshaft bearings	Three
Type of main bearings	Shimless, steel-backed, white-metal-lined
Standard main journal diameter	2.047" (52 mm)
Main journals first regrind diameter	2.027" (51.49 mm)
Main journals second regrind diameter (minimum)	2.007" (50.98 mm)
Main bearings - length	Front 1.496" (38 mm) Centre 1.496" (38 mm) Rear 1.575" (40 mm)

GENERAL DATA

TD (Continued)

Centre main bearing - end clearance	.0014" to .0037" (.035 mm to .095 mm)
Main bearing - diametrical clearance	.0008" to .003" (.020 mm to .075 mm)
Crankshaft - end thrust taken on	Centre bearing
Number of camshaft bearings	Three
Type of camshaft bearings	White metal (front) Zinc alloy (rear and centre)
Camshaft - bearing clearance	Front .0016" to .004" (.04 mm to .10 mm) Rear and centre .0018" to .0037" (.045 mm to.095 mm)
Camshaft - end thrust taken on	Front end
Camshaft - end float	.005" to .013" (.125 mm to .325 mm)
Camshaft - drive (type)	Duplex chain, 3/8" pitch, 60 pitches, endless
Valve timing markings	White chain links and "T" marks on wheels
Exhaust valve throat diameter	1.024" (25 mm)
Exhaust valve diameter	Head 1.221" (31 mm); Stem .315" (8 mm)
Inlet valve throat diameter	1.181" (30 mm)
Inlet valve diameter	Head 1.299" (33 mm); Stem .315" (8 mm)
Valve seat angle	30°
Tappet type	Hollow
Valve lift - inlet	.315" (8 mm)
Valve lift - exhaust	.315" (8 mm)
Inlet valve opens (to engine XPAG/TD2/24115)	11° before T.D.C.
Inlet valve closes	57° after B.D.C.
Exhaust valve opens	52° before B.D.C.
Exhaust valve closes	24° after T.D.C.
Inlet valve opens (from engine XPAG/TD2/24116)	5° before T.D.C.
Inlet valve closes	45° after B.D.C.
Exhaust valve opens	45° before B.D.C.
Exhaust valve closes	5° after T.D.C.
Valve spring pressure - shut	Inner 31 lb at 1.753" (14.06 kg at 4.453 cm) Outer 62 lb at 1.847" (28.12 kg at 4.692 cm) } Total 93 lb (42.18 kg)
Valve spring pressure - open	Inner 43 lb at 1.438" (19.5 kg at 3.678 cm) Outer 80 lb at 1.532" (36.29 kg at 3.89 cm) } Total 123 lb (55.79 kg)
Engines up to No XPAG/TD2/24115:	
Inlet and exhaust valve working clearance (hot)	.019" (.48 mm)
Engines from No. XPAG/TD2/24116:	
Inlet and exhaust valve working clearance (hot)	.012" (.30 mm)
Valve guides	Removable
Fuel system	
Fuel tank reserve level	Warning light on panel
Fuel delivery	S.U. electric pump
Carburettors	S.U. semi-downdraught
Carburettor needles	Standard - ES. Weak - AP. Rich - EM.

GENERAL DATA

TD (Continued)

Clutch					
Type (up to Engine No. 9407)		Borg & Beck 7 1/4" (18.4 cm) diam. Dry plate			
Type (from Engine No. 9408)		Borg & Beck 8" (20.3 cm) diam. Dry plate			
Facing		Type RYZ on 7" diameter clutch			
		Moulded type on 8" diameter clutch			
Gearbox					
Synchromesh		Second, third and top			
		Gearbox	*Overall 8/41*	*8/39*	*9/41*
Ratios	Top	1.00	5.125	4.875	4.555
	Third	1.385	7.098	6.752	6.309
	Second	2.07	10.609	10.09	9.429
	Bottom	3.5	17.938	17.06	15.942
	Reverse	3.5	17.938	17.06	15.942

Front Suspension and Steering	
Camber (static position)	Nil (tolerance ±1°)
Castor angle	2° ± 1/2° with side-members parallel to road
Toe-in	Nil
King-pin inclination	9° to 10 1/2° full bump
Track Front	47 3/8" (1.203 m), Rear 50" (1.27 m)
Turning circle	31' 3" (9.525 m)
Wheelbase	7' 10" (2.388 m)
Tyre size	5.50 - 15
Tyre pressures	18 lb per sq in (1.27 kg/cm^2)
Rear Axle	
Type of axle	Semi-floating
Type of drive	Hypoid
Ratio	8/41, 8/39 or 9/41
Adjustment	By spacers and special fixtures
Brakes	
Type	Lockheed hydraulic two-leading shoe (front) (9" dia)
Type of linings	Ferodo MR.19
Lining size - front	8.75" x 1/5" x .187" (22.22 cm x 3.81 cm x .47 cm)
Lining size - rear	8.75" x 1.5" x .187" (22.22 cm x 3.81 cm x .47 cm)
Number of rivets	12 per lining
Springs (front)	
Type	Coil
Free length	9.59" ± 1/16" (24.36 cm ± 1.58 mm)
Mean coil diameter	3.238" (8.24 cm)
Number of effective coils	7 1/2
Diameter of wire (ground)	.498" (1.27 cm)
Maximum deflection	4.24" (10.78 cm)

GENERAL DATA

TD (Continued)

Springs (rear)	
Type	Half-elliptic
Length	42" (106.7 cm)
Width	1 1/2" (3.81 cm)
Number of leaves	Seven
Thickness of leaves	7/32" (5.56 mm)
Camber (free)	4.1" (10.4 cm)
Working load	500 lb at 1/2" positive camber (226.8 kg at 12.7 mm positive camber)
Hydraulic Dampers	
Type	Girling or Armstrong
Electrical	
Distributor rotation	Counter-clockwise
Automatic advance	32° on crankshaft
Contact breaker gap (first type)	.010" to .012" (.25 mm to .30 mm)
Contact breaker gap (second type)	.014" to .016" (.36 mm to .41 mm)
Sparking plugs (engines prior to No. XPAG/TD2/22735)	Champion L.10S (std. equip.) (14 mm) KLG E.80 Lodge HN or HNP
Sparking plugs (engines from No. XPAG/TD2/22735)	Champion NA.8
Sparking plug gap	.020" to .022" (.50 mm to .56 mm)
Ignition timing	T.D.C. (full retard)
Charging system	C.V.C.
Battery	Lucas GTW9A, 51-amp at 10 hr. 12-volt. Positive earth return
Capacities	
Sump	9 pints (10.8 US pints; 5.1 litres) from Engine No. 14948 10 1/2 pints (12.6 US pints; 5.96 litres)
Gearbox	1 1/4 pints (1.5 US pints; .71 litre)
Rear axle	2 1/4 pints (2.7 US pints; 1.3 litres)
Cooling system	12 pints (14.4 US pints; 6.8 litres)
Brake fluid	1 pint (1.2 US pints; .57 litre)
Fuel tank	12 1/2 gallons (15 US gallons; 57 litres)
General Dimensions	
Length	145" (3.683 m)
Width	58 5/8" (1.489 m)
Height	53" (1.346 m)
Ground clearance	6" (15.24 cm)

GENERAL DATA

TD (Continued)

Weight (unladen)	18 1/2 cwt. (2072 lb) (941 kg)
Weight (laden and with two passengers)	22 cwt. (2464 lb) (1120 kg)
Torque Spanner Data	
Cylinder head stud nuts	500 lb/in (5.75 m/kg)
Connecting rod big-end bolts (to next split pin hole)	320 lb/in (3.7 m/kg)
Main bearing cap nuts (to next split pin hole)	750 lb/in (8.6 m/kg)
Steering wheel attachment nut	500 lb/in (5.75 m/kg)
Gudgeon pin clamp screw	400 lb/in (4.6 m/kg)

TF

All data as TD except for:-

Engine type	XPAG/TF (to Chassis No. 6500) XPEG/TF (from Chassis No. 6501)

	XPAG/TF	XPEG/TF
Capacity	1250 cc (76 1/4 cu in)	1466 cc (89.4 cu in)
Bore	66.5 mm (2.618")	72 mm (2.835")
Compression ratio	8.1 :1	8.3 : 1
RAC rating	10.97 hp	12.86 hp
BHP	57 at 5500 rpm	63 at 5000 rpm

Radiator hose top	Length 4 5/16" Diam. 1 1/2"
Crankpin minimum diameter for regrind	No regrind permissible on XPEG
Main journals regrind diameter	No regrind permissible on XPEG
Exhaust valve throat diameter	1.126" (28.6 mm)
Exhaust valve diameter	Head 1.338" (34 mm); Stem .315" (8 mm)
Inlet valve throat diameter	1.274" (32.6 mm)
Inlet valve diameter	Head 1.417" (36 mm); Stem .315" (8 mm)
Inlet valve opens	5° before T.D.C.
Inlet valve closes	45° after B.D.C
Exhaust valve opens	45° before B.D.C.
Exhaust valve closes	5° after T.D.C.

Valve spring pressure - shut	Inner 41.1 lb at 1.753" (18.64 kg at 4.453 cm) Outer 73.25 lb at 1.847" (33.23 kg at 4.692 cm)	Total 114.35 lb (51.87 kg)
Valve spring pressure - open	Inner 55 lb at 1.438" (24.95 kg at 3.678 cm) Outer 95 lb at 1.532" (43.09 kg at 3.89 cm)	Total 150 lb (68.04 kg)

Inlet valve opens - piston traverse	.009" (.23 mm)
Inlet valve working clearance (hot)	.012" (.30 mm)
Exhaust valve working clearance (hot)	.012" (.30 mm)
Fuel System	
Carburettor needles	Standard - GJ. Rich - HI. Weak - GL
Clutch	
Type	Borg & Beck 8" (20.3 cm) diam. Dry plate
Facing	Woven yarn type

GENERAL DATA

TF (Continued)

Front Suspension and Steering	
Camber (static position)	1° positive (tolerance ± 1°)
Track (disc wheels)	Front 47 3/8" (1.203 m). Rear 50" (1.27 m)
Track (wire wheels)	Front 48 3/16" (1.224 m). Rear 50 13/16" (1.29 m)
Rear Axle Ratio	8/39 = 4.875:1.
Alternative ratios	4.55 : 1 and 5.125 : 1
Brake linings	Ferodo DM.7
Springs (rear)	
Length	42 1/2" (107.95 cm)
Camber (free)	2.85" (7.24 cm)
Working load	397 lb at nil camber.
Electrical	
Automatic advance	25° on crankshaft at 3100 rpm
Contact breaker gap	.014" to .016" (.36 mm to .40 mm)
Sparking plug	Champion NA.8 (standard equipment) (14 mm), 3/4" reach
Battery	Lucas GTW9A2, 51-amp at 10 hours. 12-volt. Positive earth return
Capacities	
Sump	10 1/2 pints (12.6 US pints; 5.96 litres)
Cooling system	10 1/4 pints (12.3 US pints; 5.7 litres)
Fuel tank	12 gallons (14.4 US gallons; 54 litres)
General Dimensions	
Length	147" (3.74 m)
Width	59 3/4" (1.518 m)
Height	52 1/2" (1.34 m)
Turning circle RH	31' 3" (9.525 m)
Turning circle LH	31' 3" (9.525 m)
Weight (unladen)	17 cwt. 1 qr (1932 lb) (878 kg)
Torque Spanner Data	
Cylinder head stud nuts	600 lb/in (6.9 kg/m) (up to Engine No. XPAG/TF/31727) 500 lb/in (5.75 kg/m) (from Engine No. XPEG/TF/31728)

INDEX

A MUST FOR ALL MG OWNERS

SU CARBURETTERS - TUNING TIPS & TECHNIQUES

The SU carburetter is one of the unsung success stories of the British motor industry, appearing over and over again as part of the underbonnet scene in vehicles as different as the Mini and Rolls-Royce. For many years, twin or triple SU's were standard wear on MGs. Austin-Healeys, Jaguar E-Types and many other models, both sports and saloons, which are now widely sought-after and are regularly being restored and maintained.

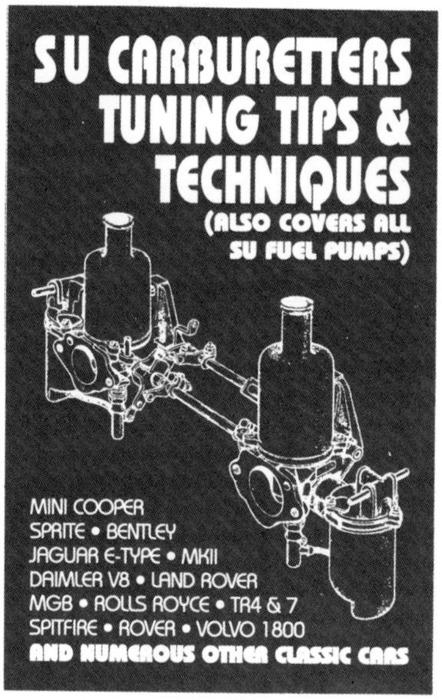

This comprehensive guide to SU tuning now appears in its 5th edition, and is as much in demand today as it was when first published. Beginning with an explanatory chapter on 'design and function', the book covers all aspects of tuning for the DIY owner and racer written in easy to follow language throughout and illustrated with a helpful variety of charts, drawings and detailed exploded views.

Concluding the book are tables of fuel systems application details (arranged alphabetically by make) and jet needle identification. Also dealt with are all SU mechanical and electrical fuel pumps.

From specialist booksellers or, in case of difficulty, direct from the distributors:
Brooklands Books Ltd., PO Box 146, Cobham, Surrey KT11 1LG, England. Phone: 01932 865051

Buy a car made in the United Kingdom.

Bigger, faster, more indomitable still — this new M.G. Midget Series T! Increased to 10 h.p. Terrifically strong chassis. Hydraulic braking on a par with its speed. A car with a racing pedigree, built to the last bolt as a pukka sports car. £222 ex works, in two-seater form only. Dunlop: Triplex.

THE M.G. CAR COMPANY LIMITED · ABINGDON-ON-THAMES · BERKSHIRE

MIDGET MG SERIES 'T'

SAFETY FAST!

Of course, the novice or the nervous *can* drive an M.G. Midget with absolute safety. But like any genuine sports car, the Midget keeps its finest performance for those who handle it with the verve and dash it deserves. The man who takes his driving seriously, as an art and a science, will understand this. Within the first mile he will know that he is controlling a car with a brilliant background of racing success. £222 ex works, in two-seater form only. Dunlops. Triplex.

Buy a car made in the United Kingdom.

FRAME: Construction of box frame, showing rubber engine mountings both front and rear.

REAR TANK: 15-gallon rear petrol tank, with quick filler-cap. Extra sturdy spare wheel bracket mounted on the chassis.

THE M.G. CAR COMPANY LIMITED . ABINGDON-ON-THAMES . BERKSHIRE
SOLE EXPORTERS—M.I.E. LTD . COWLEY . OXFORD . ENGLAND ● *M.G. owners read 'The Sports Car'*

OFFICIAL TECHNICAL BOOKS

Brooklands Technical Books has been formed to supply owners, restorers and professional repairers with official factory literature.

Workshop Manuals

Model	**Original Part No.**
Midget TC (instruction manual)	
Midget TD / TF	AKD580A
MG M to TF 1500 (Blower)	XO17
MGA 1500, 1600 & Mk. 2 (SC)	AKD600D
MGA Twin Cam	AKD926B
Midget Mk. 1, 2 & 3 & A-H Sprite	AKD4021
Midget 1500	AKM4071B
MGB & MGB GT	AKD3259 & AKD4957
MGB GT V8 Supp.	AKD8468
MGC	AKD7133/2
MGF 1995-2001	RCL0051ENG BB
MG TF 2002-2005	RCL0493(2)ENG BB
Rover 25 & MG ZR 1999-2005	RCL0534ENG BB
Rover 75 & MG ZT 1999-2005	RCL0536ENG BB

Electrical Manual	**Original Part No.**
MGF 1996-2000 MY	RCL 0342 ENG & RCL 0341 ENG

Parts Catalogues

Model	**Original Part No.**
MGA 1500 (HC)	AKD1055

Service Book	**Original Part No.**
Midget TF & TF 1500	AKD210A

Owners Handbooks

Model	**Original Part No.**
Midget TD	
MG Midget TF & TF 1500	AKD658A
MGA 1500	AKD598G
MGA 1600	AKD1172C
MGA 1600 Mk. 2	AKD1958A
Midget Mk. 3 (pub. '73)	AKD7596
Midget Mk. 3 (pub. '78)	AKM3229
Midget Mk. 3 (US) (pub. '71)	AKD7883
Midget Mk. 3 (US) (pub. '76)	AKM3436
Midget Mk. 3 (US) (pub. '79)	AKM4386
MGB Tourer (pub. '65)	AKD3900C
MGB Tourer & GT (pub. '69)	AKD3900J
MGB Tourer & GT (pub. '74)	AKD7598
MGB Tourer & GT (pub. '76)	AKM3661
MGB GT V8	AKD8423
MGB Tourer & GT (US) (pub. '68)	AKD7059B
MGB Tourer & GT (US) (pub. '71)	AKD7881
MGB Tourer & GT (US) (pub. '73)	AKD8155
MGB Tourer (US) (pub. '75)	AKM3286
MGB (US) (pub. '79)	AKM8098
MGB Tourer & GT Tuning	CAKD4034L
MGB Tuning (1800cc)	AKD4034
MGC	AKD4887B

ALSO AVAILABLE: 192 page 'Glovebox' size owners' workshop manuals:

MGA & MGB & GT 1955-68
MG Sprite & Midget 1, 2, 3, 1500 1958-80
MGB & GT 1968-81

Note: SC - Soft Cover HC - Hard Cover

From MG specialists or, in case of difficulty, direct from the distributors:

Brooklands Books Ltd., PO Box 146, Cobham, Surrey, KT11 1LG, England
Phone: 01932 865051 Fax: 01932 868803
e-mail sales@brooklands-books.com www.brooklands-books.com
Brooklands Books Australia, 3/37-39 Green Street, Banksmeadow, NSW 2019, Australia
Phone: 2 9695 7055 Fax: 2 9695 7355 www.renniks.com
Car Tech, 39966 Grand Avenue, North Branch, MN 55056, USA
Phone: 800 551 4754 & 651 277 1200 Fax: 651 277 1203 www.cartechbooks.com
Motorbooks International, P.O. Box 1, Osceola, Wisconsin 54020, USA
Phone: 800 826 6600 & 715 294 3345 www.motorbooks.com

Brooklands MG 'Road Test' & Restoration Titles

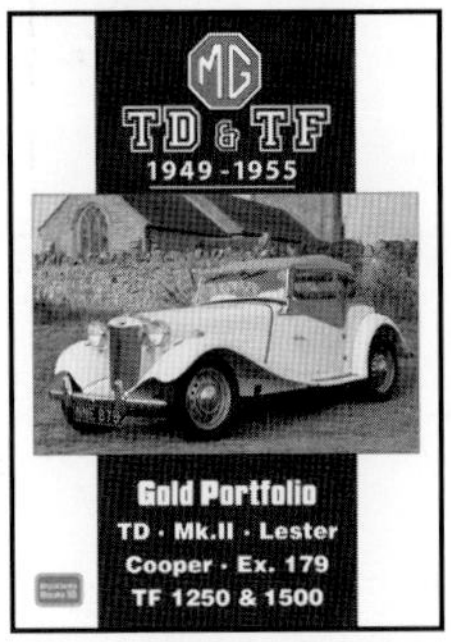

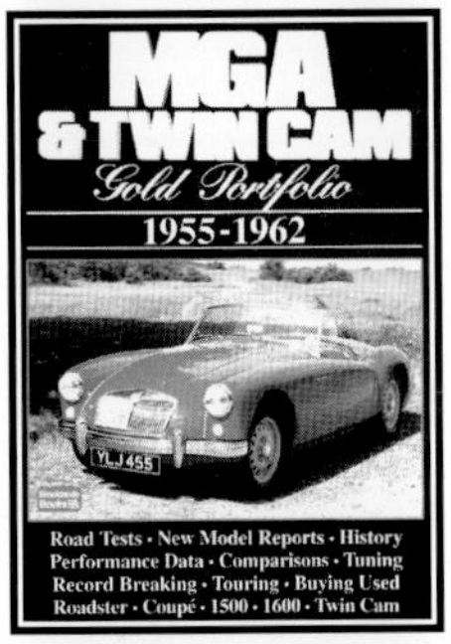

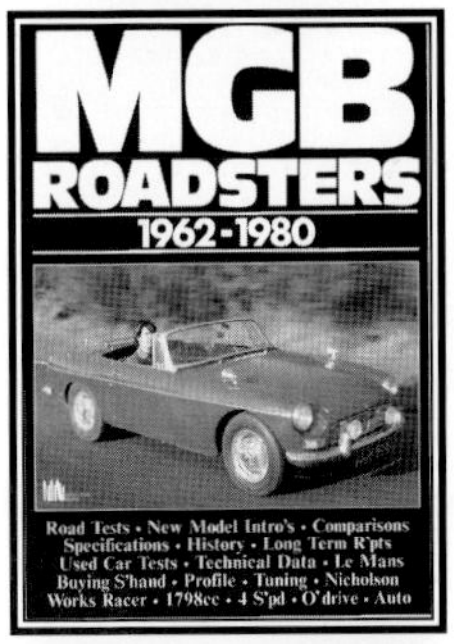

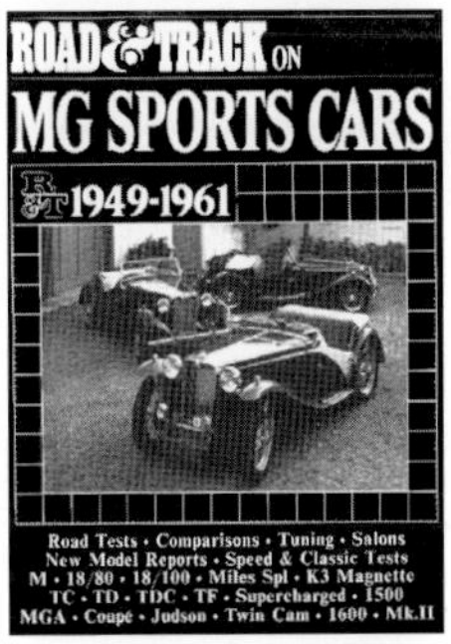

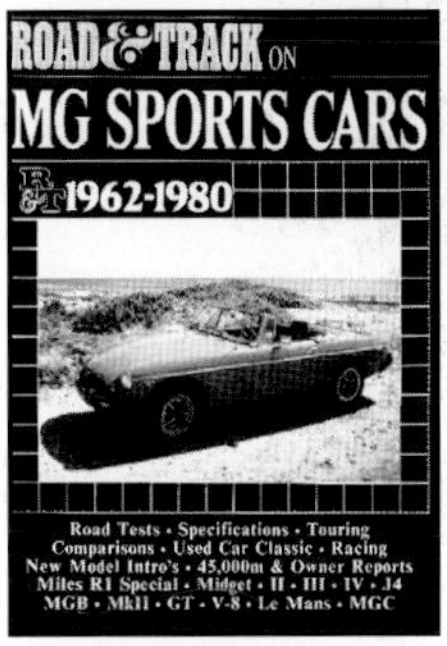

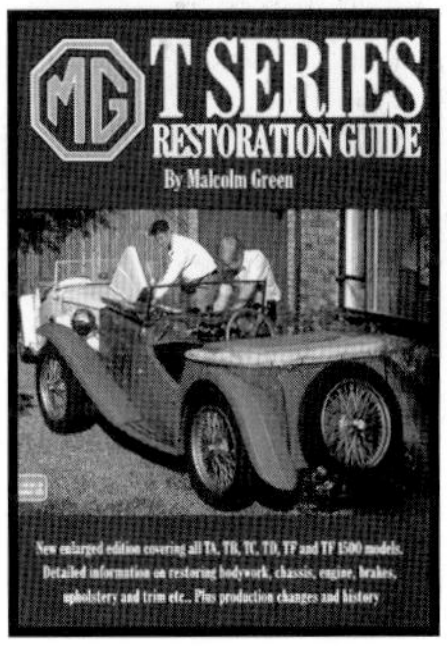

From MG specialists or, in case of difficulty, direct from the distributors:

Brooklands Books Ltd., PO Box 146, Cobham, Surrey, KT11 1LG, England.
Phone: 01932 865051 Fax: 01932 868803
E-Mail: sales@brooklands-books.com www.brooklands-books.com
Brooklands Books Australia, 3/37-39 Green Street, Banksmeadow, NSW 2019, Australia.
Telephone: 2 9695 7055 Fax: 2 9695 7355
CarTech, 39966 Grand Avenue, North Branch, MN 55056, USA.
Phone: 800 551 4754 & 651 277 1200 Fax: 651 277 1203

www.brooklands-books.com

Brooklands Books

ROAD TEST SERIES

Abarth Road Test Portfolio 1950-1971
AC Cars 1904-2009
AC & Cobra 1962-2011
Alfa Romeo Giulietta Gold Portfolio 1954-1965
Alfa Romeo Giulia Berlina Lim. Edit. Extra 1962-76
Alfa Romeo Giulia Coupes Lim. Edit. Ultra 1973-76
Alfa Romeo Alfasud 1972-1984
Alfa Romeo Spider Ultimate Portfolio 1966-1994
Alfa Romeo Spider & GTV Perf. Port. 1995-2005
Allard Limited Edition Ultra
Alpine Renault Ultimate Portfolio 1958-1995
Alvis Gold Portfolio 1919-1967
AMC Rambler Limited Edition Extra 1956-1969
AMX & Javelin Gold Portfolio 1968-1974
Armstrong Siddeley Gold Portfolio 1945-1960
Aston Martin Gold Portfolio 1921-1947
Aston Martin Ultimate Portfolio 1948-1968
Aston Martin Ultimate Portfolio 1968-1980
Aston Martin Ultimate Portfolio 1981-1993
Aston Martin Ultimate Portfolio 1994-2006
Auburn Cord Duesenberg - A Brooklands Portfolio
Audi Quattro Gold Portfolio 1980-1991
Audi TT Performance Portfolio 1998-2006
Austin Seven & Ten Road Test Portfolio
Austin-Healey 100 & 100/6 Gold Port. 1952-1959
Austin-Healey 3000 Ultimate Portfolio 1959-1967
Austin-Healey Sprite Gold Portfolio 1958-1971
Bentley & Rolls-Royce Portfolio 1990-2002
BMW 6 & 8 Cyl. Cars Limited Edition 1935-1960
BMW 700 Limited Edition 1959-1965
BMW 1600 Collection No. 1 1966-1981
BMW 2002 Ultimate Portfolio 1968-1976
BMW 6 Cylinder Coupes & Saloons Gold P. 1969-1976
BMW 316, 318, 320 (4 cyl.) Gold Port. 1975-1990
BMW 320, 323, 325 (6 cyl.) Gold Port. 1977-1990
BMW 3 Series Gold Portfolio 1991-1997
BMW M3 Ultimate Portfolio 1986-2006
BMW M5 Gold Portfolio 1980-2003
BMW 5 Series Gold Portfolio 1988-1995
BMW 6 Series Ultimate Portfolio 1976-1989
BMW 7 Series Performance Portfolio 1977-1986
BMW 7 Series Performance Portfolio 1986-1993
BMW 8 Series Performance Portfolio
BMW X5 Limited Edition Extra 1999-2006
BMW Alpina Performance Portfolio 1967-1987
BMW Alpina Performance Portfolio 1988-1998
BMW Z3, M Coupe & M Roadster Gold Port. 1996-02
Bond Cars 1949-1974 Road Test Portfolio
Borgward Isabella Limited Edition
Bristol Cars Portfolio
Bugatti Type 10 to Type 251 Road Test Portfolio
Buick Performance Portfolio 1947-1962
Buick Muscle Portfolio 1963-1973
Buick Riviera Performance Portfolio 1963-1978
Cadillac Performance Portfolio 1948-1958
Cadillac Performance Portfolio 1959-1966
Cadillac Eldorado Performance Portfolio 1967-1978
Cadillac Allante Limited Edition Extra
Caterham Seven Road Test Portfolio 1974-1999
Caterham Seven Road Test Portfolio 2000-2010
Checker Automobiles Road Test Portfolio
Impala & SS Muscle Portfolio 1958-1972
Corvair Performance Portfolio 1959-1969
El Camino & SS Muscle Portfolio 1959-1987
Chevy II & Nova SS Gold Portfolio 1962-1974
Chevelle & SS Gold Portfolio 1964-1972
Camaro Muscle Portfolio 1967-1973
Blazer & Jimmy Limited Edition Extra 1969-1982
Blazer & Jimmy Limited Edition Extra 1983-1994
Camaro Performance Portfolio 1993-2000
Chevrolet Caprice Road Test Portfolio 1965-1990
Chevrolet Corvette Gold Portfolio 1953-1962
Chevrolet Corvette Sting Ray Gold Port. 1963-1967
Chevrolet Corvette Gold Portfolio 1968-1977
Chrysler Imperial Gold Portfolio 1951-1975
Valiant 1960-1962
PT Cruiser Performance Portfolio
Citroen Traction Avant Limited Edition Premier
Citroen 2CV Ultimate Portfolio 1948-1990
Citroen DS & ID 1955-1975
Citroen DS & ID Gold Portfolio 1955-1975
Citroen SM Limited Edition Extra 1970-1975
Citroen CX Road Test Portfolio
Shelby Cobra Gold Portfolio 1962-1969
Crosley & Crosley Specials Limited Edition
Cunningham Automobiles 1951-1955
Damlier SP250 Dart & V-8 2.5 litre 250 RTP 1959-69
Datsun Roadsters Performance Portfolio 1960-71
Datsun 240Z & 260Z Road Test Portfolio
Datsun 280Z & 280ZX Road Test Portfolio
Delahaye Road Test Portfolio
Delage Road Test Portfolio
DeLorean Gold Portfolio 1977-1995
De Soto Limited Edition 1952-1960
Dodge Limited Edition 1949-1959
Dodge Dart Limited Edition Extra 1960-1976
Dodge Muscle Portfolio 1964-1971
Charger Muscle Portfolio 1966-1974
Elva Sports Racers Road Test Portfolio
ERA Gold Portfolio 1934-1994
Facel Vega Limited Edition Extra 1954-1964
Ferrari Road Cars 1946-1956 Road Test Portfolio
Ferrari Dino Limited Edition Extra 1965-1974
Ferrari 308 & Mondial Ultimate Portfolio 1975-85
Ferrari 328 348 Mondial Ultimate Portfolio 1986-94
Ferrari F355 & 360 Gold Portfolio 1995-2004
Fiat 500 1936-1972 Road Test Portfolio
Fiat Dino Road Test Portfolio
Fiat 124 Spider Performance Portfolio 1966-1985
Fiat Abarth 1972-1987 Road Test Portfolio
Fiat X1/9 Gold Portfolio 1973-1989
Fiat Barchetta Road Test Portfolio
Ford Zephyr, Zodiac, Executive Mk. III & IV 1962-1971
High Performance Capris Gold Portfolio 1969-1987
Capri Muscle Portfolio 1974-1987
High Performance Fiestas 1979-1991
Ford Escort RS & Mexico Limited Edition 1970-1979
High Performance Escorts Mk. II 1975-1980
High Performance Escorts 1980-1985
High Performance Escorts 1985-1990
Ford Thunderbird Performance Portfolio 1955-1957
Ford Thunderbird Performance Portfolio 1958-1963
Ford Thunderbird Performance Portfolio 1964-1976
Ford Fairlane Performance Portfolio 1955-1970
Ford Ranchero Muscle Portfolio 1957-1979
Ford Galaxie & LTD Gold Portfolio 1960-1976
Falcon Performance Portfolio 1960-1970
Ford GT40 & GT Ultimate Portfolio 1964-2006
Ford Torino Performance Portfolio 1968-1974
Ford Bronco 4x4 Performance Portfolio 1966-1977
Ford Bronco 1978-1988
Shelby Mustang Ultimate Portfolio 1965-1970
Mustang Muscle Portfolio 1967-1973
High Performance Mustang IIs 1974-1978
Mustang 5.0L Muscle Portfolio 1982-1993
Ginetta Cars Limited Edition Ultra 1958-2007
Goggomobil Limited Edition
Hispano-Suiza Road Test Portfolio
Honda - Acura NSX Ultimate Portfolio 1989-2005
Honda CRX 1983-1987
Honda S2000 Performance Portfolio 1999-2008
Hudson Performance Portfolio 1946-1957
International Scout Gold Portfolio 1961-1980
Isetta Gold Portfolio 1953-1964
ISO & Bizzarrini Limited Edition Ultra 1962-1974
Jaguar and SS Gold Portfolio 1931-1951
Jaguar XK120 • 140 • 150 Gold Portfolio 1948-1960
Jaguar Mk 7, 8, 10 & 420 Road Test Portfolio
Jaguar Mk 1 & Mk 2 1955-1969 Road Test Portfolio
Jaguar C-Type & D-Type Gold Portfolio 1951-1960
Jaguar E-Type Ultimate Portfolio 1961-1975
Jaguar XJ6 Series I & II Gold Portfolio 1968-1979
Jaguar XJ6 Series III Perf. Portfolio 1979-1986
Jaguar XJS Gold Portfolio 1975-1988
Jaguar XJ-S V12 Ultimate Portfolio 1988-1996
Jaguar XK8 & XKR Performance Portfolio 1996-2005
Jeep CJ-5 Limited Edition 1960-1975
Jeep CJ-5 & CJ-7 4x4 Perf. Portfolio 1976-1986
Jeep Wagoneer Performance Portfolio 1963-1991
Jeep J-Series Pickups 1970-1982
Jeepster & Commando Limited Edition 1967-1973
Jeep Cherokee & Comanche Pickups P. P. 1984-91
Jeep Wrangler 4x4 Performance Portfolio 1987-99
Jeep Cherokee & Grand Cherokee 4x4 P. P. 1992-98
Jensen Road Test Portfolio 1934-1965
Jensen Interceptor Ultimate Portfolio 1966-1992
Jensen - Healey Road Test Portfolio 1972-1976
Kaiser - Frazer Limited Edition 1946-1955
Lagonda Gold Portfolio 1919-1964
Lancia Aurelia & Flaminia Gold Portfolio 1950-1970
Lancia Fulvia Gold Portfolio 1963-1976
Lancia Beta Gold Portfolio 1972-1984
Lancia Montecarlo & Scorpion 1975-1982 RTP
Lancia Stratos Limited Edition Extra
Lancia Delta & integrale Ultimate Portfolio
Land Rover Series I, II & IIA Gold Portfolio 1948-71
Land Rover Series III 4x4 Perf. Portfolio 1971-1985
Land Rover 90 110 Defender Gold Portfolio 1983-94
Land Rover Discovery Perf. Port. 1989-2000
Lamborghini Performance Portfolio 1964-1976
Lamborghini Performance Portfolio 1977-1989
Lamborghini Gold Portfolio 1990-2004
Lincoln Gold Portfolio 1949-1960
Lincoln Continental Performance Portfolio 1961-1969
Lincoln Continental 1969-1976
Lola Sports Racing Cars Lim. Ed. Premier 1958-1985
Lotus Sports Racers Portfolio - covering 1951-1965
Lotus Seven Gold Portfolio 1957-1973
Lotus Elite Limited Edition 1957-1964
Lotus Elan Ultimate Portfolio 1962-1974
Lotus Cortina Road Test Portfolio
Lotus Elite & Exlat Road Test Portfolio 1984-1972
Lotus Excel Road Test Portfolio
Lotus Europa Gold Portfolio 1966-1975
Lotus Elise & Exige Gold Portfolio 1995-2005
Marcos Coupés & Spyders Gold Portfolio 1960-1997
Maserati Cars Performance Portfolio 1957-1970
Maserati Cars Performance Portfolio 1971-1982
Maserati Cars Performance Portfolio 1982-1998
Maserati Cars Ultimate Portfolio 1999-2007
Mazda Miata MX-5 Performance Portfolio 1989-1997
Mazda Miata MX-5 Performance Portfolio 1998-2005
Mazda Miata MX-5 Takes On The Competition
McLaren F1 • GTR • LM Sportscar Perf. Portfolio
Mercedes-Benz 1925-1939 - A Portfolio
Mercedes 190SL 300SL 300SLR - A Portfolio
Mercedes S & 600 Limited Edition Extra 1965-1972
Mercedes S Class 1972-1979
Mercedes S Class Limited Edition Extra 1980-1991
Mercedes 230SL 250SL 280SL Ultimate Port. 1963-71
Mercedes-Benz SLs & SLCs Ultimate Port. 1971-89
Mercedes SLs Performance Portfolio 1989-1994
Mercedes E-Class W124 Road Test Portfolio 1986-1995
Mercedes G-Wagen Gold Portfolio 1981-2005
Mercedes 190 Limited Edition Extra 1983-1993
Mercedes AMG Gold Portfolio 1983-1999
Mercedes AMG Ultimate Portfolio 2000-2006
Mercury Gold Portfolio 1947-1966
Mercury Comet & Cyclone Lim. Edit. Extra 1960-75
Cougar Muscle Portfolio 1967-1973
Messerschmitt Gold Portfolio 1954-1964
MG Gold Portfolio 1929-1939
MG TA & TC Gold Portfolio 1936-1949
MG TD & TF Gold Portfolio 1949-1955
MG Y Type & Magnette Road Test Portfolio
MGA & Twin Cam Gold Portfolio 1955-1962
MG Midget Road Test Portfolio 1961-1979
MGB Roadsters 1962-1980
MGB MGC & V8 Gold Portfolio 1962-1980
MGC & MGB GT V8 Limited Edition
MGF & TF Performance Portfolio 1995-2005
Mini Gold Portfolio 1959-1969
Mini Gold Portfolio 1969-1980
Mini Gold Portfolio 1981-1997
High Performance Minis Gold Portfolio 1960-1973
Mini Cooper Gold Portfolio 1961-1971
Mini Moke Ultimate Portfolio 1964-1994
Mini Performance Portfolio 2001-2006
Starion & Conquest Performance Portfolio 1982-90
Mitsubishi 3000GT & Dodge Stealth P.P. 1990-99
Morgan Three-Wheelers Ultimate Portfolio 1909-52
Morgan Four-Wheelers Ultimate Portfolio 1936-67
Morgan Ultimate Portfolio 1968-1990
Morgan Ultimate Portfolio 1991-2009
Morris Minor MM & Series II Road Test Portfolio
Morris Minor 1000 Road Test Portfolio
Nash Limited Edition Extra 1949-1957
Nash-Austin Metropolitan Gold Portfolio 1954-1962
Nissan Skyline GT-R Ultimate Portfolio 1969-2010
Noble Sports Cars Road Test Portfolio
NSU Ro80 Limited Edition
Oldsmobile 1948-1963 Limited Edition Premier
Oldsmobile Muscle Portfolio 1964-1971
Cutlass & 4-4-2 Muscle Portfolio 1964-1974
Opel GT Ultimate Portfolio 1968-1973
Opel Manta Limited Edition 1970-1975
Packard Automobiles 1920-1958 - A Portfolio
Pantera Ultimate Portfolio 1970-1995
Panther Gold Portfolio 1972-1990
Plymouth Limited Edition 1950-1960
Plymouth Fury Limited Edition Extra 1956-1976
Barracuda Muscle Portfolio 1964-1974
Plymouth Muscle Portfolio 1964-1971
Pontiac 1946- 1963 Limited Edition Premier
High Performance Firebirds 1982-1988
Firebird & Trans Am Performance Portfolio 1993-00
Pontiac Fiero Performance Portfolio 1984-1988
Porsche Sports Racing Cars UP 1952-1968
Porsche 917 • 935 • 956 • 962 Gold Portfolio
Porsche 912 Limited Edition Extra
Porsche 365 Ultimate Portfolio 1952-1965
Porsche 911 1965-1969
Porsche 911 1973-1977
Porsche 911 SC & Turbo Gold Portfolio 1978-1983
Porsche 911 Carrera & Turbo Gold Port. 1984-1989
Porsche 911 Ultimate Portfolio 1990-1997
Porsche 911 Takes On The Competition 1990-1997
Porsche 911 Ultimate Portfolio 1998-2004
Porsche 914 Ultimate Portfolio
Porsche 924 Gold Portfolio 1975-1988
Porsche 928 Gold Portfolio 1977-1995
Porsche 928 Takes On The Competition
Porsche 944 Ultimate Portfolio
Porsche 968 Limited Edition Extra
Porsche Boxster Ultimate Portfolio 1996-2004
Railton & Brough Superior Gold Portfolio 1933-1950
Range Rover Gold Portfolio 1970-1985
Range Rover Gold Portfolio 1985-1995
Range Rover Performance Portfolio 1995-2001
Range Rover Takes on the Competition
Riley RM Series Pathfinder 2.6 A Brooklands Port.
Rolls-Royce Silver Clould & Bentley S Ultimate Port.
Rolls-Royce Silver Shadow Ultimate Port. 1965-80
Rover P4 1949-1959
Rover 2000 & 2200 1963-1977
Studebaker Ultimate Portfolio 1946-1966
Studebaker Hawks & Larks Lim. Edit. Premier 1956-66
Avanti Limited Edition Extra 1962-1991
Subaru Impreza Turbo Limited Edition Extra 1994-00
Subaru Impreza WRX Performance Port. 2001-2005
Sunbeam Alpine Limited Edition Extra 1959-1968
Sunbeam Tiger Limited Edition Extra 1964-1967
Suzuki SJ Gold Portfolio 1971-1997
Vitara, Sidekick & Geo Tracker Perf. Port. 1988-1997
Tatra Cars Road Test Portfolio
Toyota Land Cruiser Gold Portfolio 1956-1987
Toyota Land Cruiser 1988-1997
Toyota Supra Performance Portfolio 1982-1998
Toyota MR2 Road Test Portfolio 1984-1989
Toyota MR2 Road Test Portfolio 1990-1999
Toyota MR2 Road Test Portfolio 2000-2007
Toyota MR2 Takes On The Competition
Triumph TR2 & TR3 Gold Portfolio 1952-1961
Triumph TR4, TR5, TR250 1961-1968
Triumph TR6 Road Test Portfolio
Triumph Herald 1959-1971
Triumph Vitesse 1962-1971
Triumph 2000, 2.5, 2500 1963-1977
Triumph GT6 Gold Portfolio 1966-1974
Triumph Stag Road Test Portfolio
TVR Limited Edition Ultra 1958-1985
TVR Performance Portfolio 1986-1994
TVR Performance Portfolio 1995-2000
TVR Performance Portfolio 2000-2005
VW Beetle Gold Portfolio 1935-1967
VW Beetle Gold Portfolio 1968-1991
VW Bus Camper Van Perf. Portfolio 1954-1967
VW Bus Camper Van Perf. Portfolio 1968-1979
VW Bus Camper Van Perf. Portfolio 1979-1991
VW Karmann Ghia Gold Portfolio 1955-1974
VW Scirocco 1974-1981
VW Golf GTI Limited Edition Extra 1976-1991
VW Golf Cabriolet 1979-2002 Road Test Portfolio
VW Corrado Limited Edition Premier 1989-1995
Volvo PV444 & PV544 Perf. Portfolio 1945-1965
Volvo 120 Amazon Ultimate Portfolio
Volvo 1800 Ultimate Portfolio 1960-1973
Volvo 140 & 160 Series Gold Portfolio 1966-1975
Forty Years of Selling Volvo
Westfield Performance Portfolio 1982-2004

MILITARY VEHICLES

Complete WW2 Military Jeep Manual
Dodge WW2 Military Portfolio 1940-1945
German Military Equipment WW2
Hail To The Jeep
Combat Land Rover Portfolio No. 1
Land Rover Military Portfolio
Military & Civilian Amphibians 1940-1990
Off Road Jeeps Civilian & Military 1944-1971
US Military Vehicles 1941-1945
Standard Military Motor Vehicles-TM9-2800 (WW2)
VW Kubelwagen Military Portfolio 1940-1990
WW2 Allied Vehicles Military Portfolio 1939-1945
WW2 Jeep Military Portfolio 1941-1945

RACING & THE LAND SPEED RECORD

The Land Speed Record 1898-1919
The Land Speed Record 1920-1929
The Land Speed Record 1930-1939
The Land Speed Record 1940-1962
The Land Speed Record 1963-1999
Can-Am Racing 1966-1969
Can-Am Racing 1970-1974
Can-Am Racing Cars 1966-1974
The Carrera Panamericana Mexico - 1950-1954
Le Mans - The Bentley & Alfa Years - 1923-1939
Le Mans - The Jaguar Years - 1949-1957
Le Mans - The Ferrari Years - 1958-1965
Le Mans - The Ford & Matra Years - 1966-1974
Le Mans - The Porsche Years - 1975-1982
Le Mans - The Porsche & Jaguar Years - 1983-91
Le Mans - The Porsche & Peugeot Years - 1992-99
Mille Miglia - The Alfa & Ferrari Years - 1927-1951
Mille Miglia - The Ferrari & Mercedes Years - 1952-57
Targa Florio - The Porsche & Ferrari Years - 1955-1964
Targa Florio - The Porsche Years - 1965-1973

RESTORATION & GUIDE SERIES

BMW 2002 - A Comprehensive Guide
BMW 02 Restoration Guide
BMW E30 - 3 Series Restoration Bible
Classic Camaro Restoration
Engine Swapping Tips & Techniques
Ferrari Life Buyer's Portfolio
Land Rover Restoration Portfolio
PC on Land Rover Series I Restoration
Lotus Elan Restoration Guide
MG T Series Restoration Guide
MGA Restoration Guide
PC on MGB Restoration
Mustang Restoration Tips & Techniques
Practical Gas Flow
Restoring Sprites & Midgets an Enthusiast's Guide
SU Carburetters Tuning Tips & Techniques
PC on Sunbeam Rapier Restoration
The Great Classic Muscle Cars Compared
Weber Carburettors Tuning Tips and Techniques

ROAD & TRACK SERIES

Road & Track on Aston Martin 1962-1990
Road & Track on Austin Healey 1953-1970
Road & Track on BMW Cars 1966-1974
Road & Track BMW M Series Portfolio 1979-2002
R & T BMW Z3, M Coupe & M Roadster Port. 96-02
R & T Camaro & Firebird Portfolio 1993-2002
Road & Track on Corvette 1968-1982
Road & Track on Corvette 1982-1986
Road & Track on Corvette 1986-1990
Road & Track Corvette Portfolio 1997-2002
Road & Track Dodge Viper Portfolio 1992-2002
Road & Track on Ferrari 1975-1981
Road & Track on Ferrari 1984-1988
Road & Track Ferrari V-12 Portfolio 1992-2002
Road & Track Ferrari F355 360 F430 Portfolio 95-06
Road & Track on Jaguar 1950-1960
Road & Track on Jaguar 1961-1968
Road & Track on Jaguar 1968-1974
R & T Jaguar XJ-S - XK8 - XKR Portfolio 1975-2003
Road & Track MX-5 Miata Portfolio 1989-2002
Road & Track on Mercedes 1952-1962
R & T Mercedes SL - SLK - CLK Portfolio 1990-2003
Road & Track on MG Sports Cars 1949-1961
Road & Track on MG Sports Cars 1962-1980
Road & Track Mustang Portfolio 1994-2002
Road & Track Nissan 300ZX & 350Z Portfolio 1984-03
Road & Track on Porsche 1951-1967
Road & Track on Porsche 1972-1975
Road & Track on Porsche 1975-1978
Road & Track on Porsche 1979-1982
Road & Track Porsche 928 Portfolio 1977-1994
Road & Track Porsche 911 Portfolio 1990-1997
R & T on Rolls Royce & Bentley 1950-1965
R & T on Rolls Royce & Bentley 1966-1984
R & T on Triumph Sports Cars 1967-1974
R & T on Triumph Sports Cars 1974-1982
Road & Track on Volkswagen 1951-1968
Road & Track on Volkswagen 1968-1978
Road & Track on Volvo 1957-1974
Road & Track on Volvo 1977-1994
Road & Track - Best of PS
Road & Track - Peter Egan Side Glances 1983-1992
Road & Track - Peter Egan Side Glances 1992-1997
Road & Track - Peter Egan Side Glances 1998-2002
Road & Track - Peter Egan Side Glances 2002-2006

CAR AND DRIVER SERIES

Car and Driver on BMW 1957-1977
Car and Driver on Ferrari 1955-1962
Car and Driver on Ferrari 1963-1975
Car and Driver on Porsche 1955-1962
Car and Driver on Porsche 1963-1970
Car and Driver on Porsche 1970-1976
Car and Driver on Porsche 1977-1981
Car and Driver on Porsche 1982-1986
Car and Driver on Volvo 1955-1976

HOT ROD 'ENGINE' SERIES

Chevy 265 & 283
Chevy 302 & 327
Chevy 348 & 409
Chevy 396 & 427
Chevy 454 thru 512
Chevy Monster Big Blocks
Chrysler Hemi
Chrysler 273, 318, 340 & 360
Chrysler 361, 383, 400, 413, 426 & 440
Ford 289, 302, Boss 302 & 351W
Ford 351C & Boss 351
Ford Small Block
Ford Big Block

MOTORCYCLES

To see our range of over 70 titles visit
www.brooklands-books.com